AEROSPACE STRUCTURES
An Introduction

Kruger Brentt
Publishers

AEROSPACE STRUCTURES
An Introduction

Flower Fang
Alice Moss

Kruger Brentt
Publishers

2024

Kruger Brentt Publishers UK. LTD.
Company Number 9728962

Regd. Office: 68 St Margarets Road, Edgware, Middlesex HA8 9UU

ISBN: 978-1-78715-154-3

For information on all our publications visit our website at http://krugerbrentt.com/

PREFACE

Aerospace structures differ from other structures due to their high demands for performance and lightweight. Modern aerospace structures typically require the use of composite materials, advanced multifunctional materials and thin-walled constructions. An aerostructure is a component of an aircraft's airframe. This may include all or part of the fuselage, wings, or flight control surfaces. Companies that specialize in constructing these components are referred to as "aerostructures manufacturers", though many larger aerospace firms with a more diversified product portfolio also build aerostructures. Mechanical testing of the individual components or complete structure is carried out on a Universal Testing Machine. Test carried out include tensile, compression, flexure, fatigue, impact, compression after impact. Before testing the component, aerospace engineers build finite element models to simulate the reality.

Airplanes designed for civilian use are often cheaper than military aircraft. Smaller passenger airplanes are used for short distance, transcontinental transport. It is more cost efficient for airlines and there is less demand for aircraft transportation at these distances as people can, while inconvenient, drive these distances. While bigger airplanes are manufactured for intercontinental transport, so more passengers can be carried at one time, money can be saved on fuel, and airliners do not have to pay as many pilots. Cargo planes are usually built to be bigger than the average jet. They have a lot of space and large dimensions, so they can carry a lot of weight and a large volume of cargo in one trip. They have large wingspans, a very large cargo hold, and a very tall vertical fin. They are not built to accommodate passengers except for the pilots, so the use of the cargo hold is much more efficient. There does not need to be room for seats and food and bathrooms for everybody, so the companies made a design that optimizes the space in the aircraft.

The present book contains 11 chapters namely Material physics & properties; environment and durability; material types; manufacturing; aircraft & spacecraft structures; aircraft & spacecraft load; translating loads to stresses; considering strength & stiffness; design & certification; fatigue & durability and structural joints.

The book covers all the fundamental aspects for engineering students such as structural analysis, elasticity, aero elasticity and airworthiness. It emphasizes basic structural theory, which remains unchanged with the development of new materials and construction methods. It also describes the application of the elementary principles of mechanics to the analysis of aircraft structures. This book is useful for undergraduate as well as postgraduate students of aerospace and aeronautical engineering.

We are grateful to all those persons as well as various books, manuals, periodicals, magazines, journals etc. that helped in the preparation of this book. In spite of the best efforts, it is possible that some errors may have occurred into the compilation and editing of the book. Further queries, constructive suggestions and criticisms for the improvement of the book are always welcomed and shall be thankfully acknowledged.

Flower Fang

Alice Moss

CONTENTS

Contents

INTRODUCTION

Introductions into aerospace comprise the introduction into many aerospace related disciplines, and their interrelations. The major message generally is that an optimum in aerospace constitutes compromises the related disciplines. Similarly, aerospace materials and structures represent a field in which structural engineering, material science and manufacturing contribute equally, making trade-offs and compromises necessary.

This textbook is written to fill the gap between these general introductions into aviation and textbooks covering either material science, mechanics of materials or structural analyses. Where the first are deemed insufficient to cover the basic aspects of these disciplines, the latter miss the relevant interrelations between the disciplines.

Students are warned prior to reading this book; the field of aerospace structures and materials is not solely exact science or engineering. Chapters are presented that are indeed rather scientific or engineering of nature (solid material physics, and structural analysis) allowing for theories or solutions based on formulas and equations, but other chapters are more qualitative and philosophical (safety, manufacturability, availability and costing). Students should be aware that in the long end, decisions made within the field of aerospace structures and materials are often dictated by these soft considerations rather than hard core engineering. The main objective of this textbook therefore, is to create awareness and a critical mind-set to aid the student when pursuing a study in aerospace engineering.

This book forms an update of a course reader that I wrote many years ago. Publishing this book has been made possible with the help of many. In particular I would like to thank my colleague Gillian Saunders-Smits for coordinating and contributing to the process and suggestions for additions, Hilde Broekhuis for converting the reader text to the book format, editing and updating the illustrations, Calvin Rans, Urban Avsec and Katharina Ertman for assisting in finding and creating illustrations, Renée van de Watering and Cora Bijsterveld for their help with copyright, and the staff of the TU Delft Library under the coordination of

Michiel de Jong, who contributed to development of the book and assist us with the many challenges of developing an Open Text Book. I would also like to acknowledge the learners in the first run of our Massive Open Online Course "Introduction to Aerospace Structures & Materials" on edX whose critical reading eliminated many small errors. The quality of this textbook is to great extent a result from their effort and criticism, which I greatly appreciate.

CHAPTER-1

MATERIAL PHYSICS & PROPERTIES

1.1 INTRODUCTION

This chapter will discuss the elementary physics of materials related to the loads acting on a material and as a consequence their response. The response is related to physical properties of materials which can substantially differ from one material to another. This chapter highlights some of the differences in the material propertiesthat are observed in commonly used structural materials.

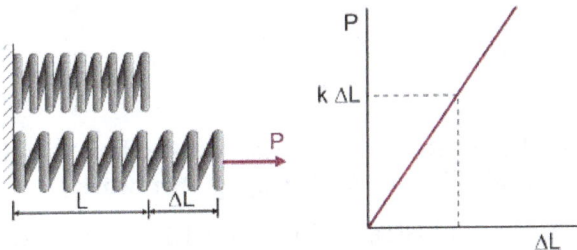

Figure 1.1 Illustration of a spring loaded with load P and its subsequent load-elongation diagram (Alderliesten, 2011, 1-1.jpg. Own Work)

When applying a load P to a spring with length L, it will elongate with ΔL Figure 1.1. This elongation relates linearly to the applied load P and is often formulated as

$$P= k\Delta L \qquad\qquad (1.1)$$

where k is called the spring constant. While loading the spring, one may observe thatthe diameter of the spring becomes smaller, the longer the spring is stretched. This loaded spring represents the elastic behaviour of materials in general when loaded uni-axially; for given load the material will elongate, while the cross-section becomesslightly smaller.

1.2 STRESS-STRAIN

The material behaviour referred to earlier is represented by other parameters than elongation and load, because the magnitude of the load for a given elongation (represented by for example a spring constant, k), depends on the shape or geometry of the material. Different geometries or original lengths of the same material will thus give different load-displacement curves, which is inconvenient when comparingmaterials.

The parameters used to evaluate the material properties are selected based on whatis called the similitude principle. To physically equate the proportional relationship between load and material response, dimensional aspects should be left out of the equation.

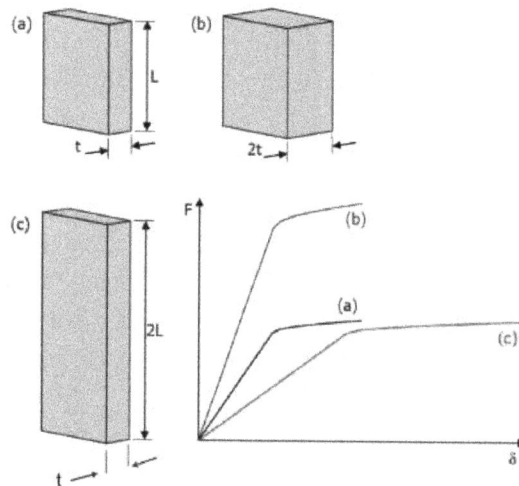

Figure1.2 Illustration of the geometry influence on the force displacement response (Alderliesten, 2011, 1-2.jpg. Own work.)

Consider the three samples illustrated in Figure 1.2. 1To elongate the samples (a) and (b) equally, the load F applied on sample (b) should be twice as large as on sample (a). In the force-displacement diagram this results in a curve for sample (b) twice as high as sample (a). For the same applied force F, sample (c) will elongate twice as much as sample (a). In the diagram this results in a curve thatis stretched twice as much as curve (a).

Although all three samples are made of the same material, the curves appear to be different. To exclude the dimensional aspects from the material's response to load, the selected parameters should be chosen 'dimensionless', i.e. independent of dimensions. Based upon curves (a) and (b), the force must be divided by the cross section of the sample, resulting in stress σ, and based upon curves (a) and (c) the elongation must be divided by the sample's length, resulting in the dimensionless strain ε.

For this reason the extension of the material is represented by strain, which is the extension normalized by its initial length according to

$$\varepsilon = \frac{\Delta L}{L} \qquad\qquad (1.2)$$

Similarly, the effect of geometry is excluded by representing the load application in terms of stresses

$$\sigma = \frac{P}{A} \qquad\qquad (1.2)$$

where A is the cross section of the material.

There are two ways to calculate the stress with equation (1.3):

⊙ Dividing the load by the original or initial cross-section, Ao. The stress is then called the engineering stress

⊙ Dividing the load by the actual cross-section A. The stress is then called the true stress

Since the actual cross-section is often not exactly known, the engineering stress is often taken for stress analysis.

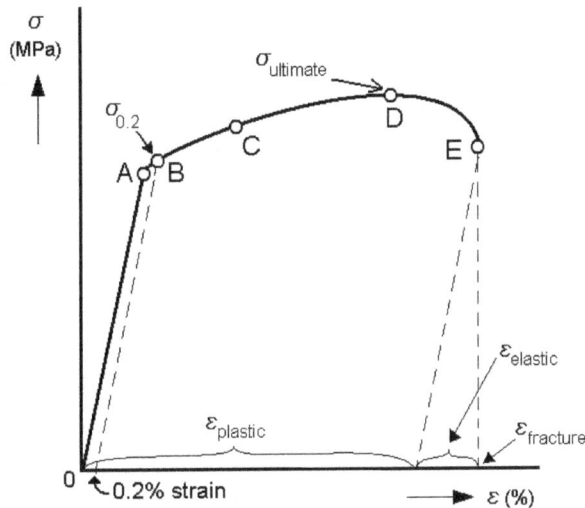

Figure 1.3 Typical stress-strain curve for a metal; initial slope is linear elastic, beyond yielding the material behaviour is plastic (TU Delft, n.d. 1-3.jpg. Own work)

The stress with equation (1.3) can be plotted against the strain calculated with equation (1.2), which for many elastic-plastic materials like metals, gives a curve as illustrated in Figure 1.3.

The initial slope of the curve is linear-elastic, which means that when unloading, the material will return to its original length and shape. Beyond a certain load, the material will permanently deform. This transition point in the stress-strain curve is called the yield point. Because often the yield point is a gradual transition from the linear elastic curve into the plastic region, it is hard to determine the yield stress exactly in an equal manner for all materials. For this reason, a common (but arbitrary) approach is to take the intersection between the stress-strain curve and the 0.2% offset of the linear elastic slope, illustrated with the dotted line in Figure 1.3.

1.3 LOADING MODES

The examples so far (i.e. elongation of spring or material) assumed a uni-axial loading mode in tension. Often compression is assumed to be identical to tension except for the sign (direction). These two loading modes are illustrated in Figure 1.4 together with the shear loading and torsional loading mode. Depending on the shape of material or structure and the load applied, the material may face either one of these four modes, or a combination of them.

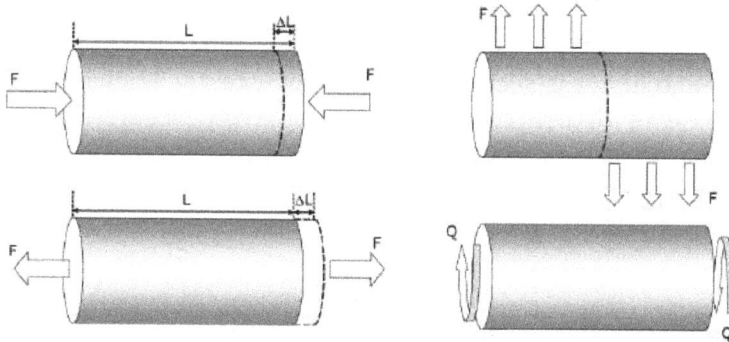

Figure 1.4 Four loading modes: compression, tension, shear and torsion (Alderliesten, 2011, 1-4.jpg. Own Work.)

1.4 ENGINEERING TERMINOLOGY

The stress-strain curve illustrated in Figure 1.3 contains terminology that requires some explanation. For that purpose, two different curves are being given in Figure 1.5.

Concerning the linear elastic part (initial slope of the curves), the slope may be either steep (high resistance against deformation) or gentle (representing low resistance against deformation). The first curve indicates a stiff material, whereas the second curve indicates a flexible material.

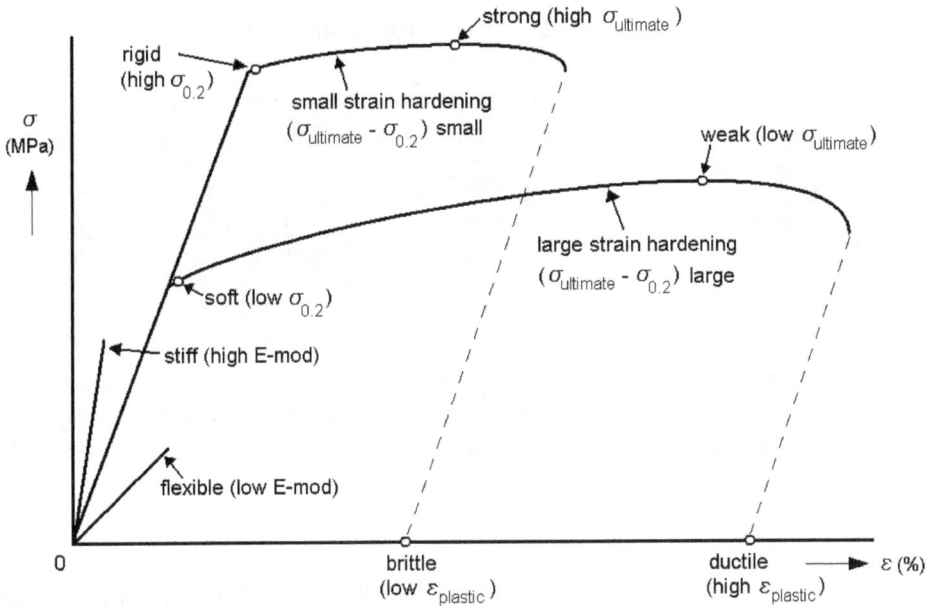

Figure 1.5: Stress-Strain Diagram which shows the difference between stiff and flexible (high or low E-modulus, respectively) material, soft and rigid (low or high yield stress, resp.) and small and large strain hardening (small or large difference between ultimate and yield stress, resp.) and weak and strong material (low or high ultimate stress, resp.) and brittle and ductile material, (low or high plastic strain resp.). (TU Delft, n.d. 1-5.jpg. Own Work)

The yield point (transition from elastic to plastic) may either be located at small values of the stress (low yield strength) or at high values of stress (high yield strength). The first transition point indicates a soft material, whereas the second indicates a rigid material.

After yielding, the curves continue to increase gradually. The hypothetical case where the material becomes fully plastic after yielding, i.e. the slope continues horizontally, is often denoted as perfect plastic. In all other cases, there is a slope that is either rather steep (large strain hardening) or gentle (small strain hardening).

The highest point in the stress strain curve is called the ultimate strength of a material. If this strength value is very high, it indicates a strong material. If the strength is low, it indicates a weak material.

Fracture occurs at the end of the curve. The elastic deformation still present causes spring back. This is illustrated by the dotted lines parallel to the initial elastic slope of the curve. The remaining deformation is plastic deformation. A small degree of plastic deformation indicates a brittle material. A high degree of plastic deformation indicates a ductile material.

Note, that there is a fundamental difference in *strength and stiffness* (see Figure 1.6).

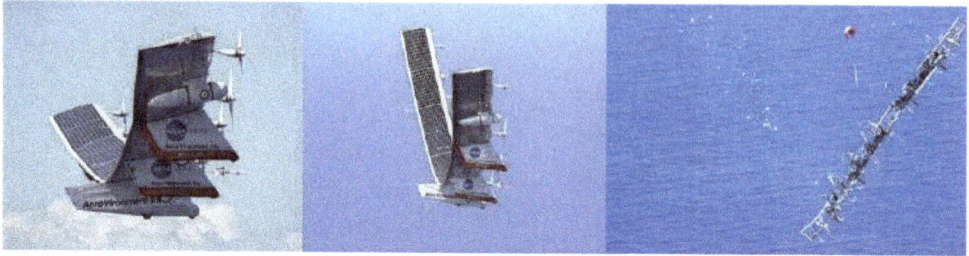

Figure 1.6 Illustration of the difference between stiffness and strength; the flexibility of the wings relates to stiffness (low stiffness gives significant wing bending), whereas strength relates to final failure of the structure. Derivative from NASA, (2003), Public Domain.

1.5 NORMAL STRESS

In the case of tension and compression (see Figure 1.4) normal stresses occur in the material. According to the sign convention, these stresses are either positive (tension) or negative (compression). Similar to the spring constant, the relation between stress and strain is characterized by a constant in the linear elastic part of the stress strain curve.

$$\sigma = E\varepsilon \qquad\qquad (1.4)$$

The constant E is called the modulus of elasticity, or the Young's modulus. The value of this Young's modulus is a characteristic value for a material; a high value indicates a stiff material, a low value a flexible material.

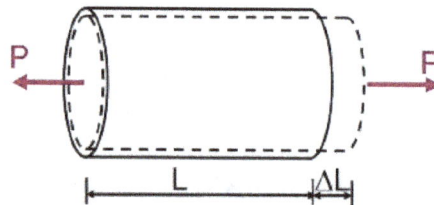

Figure 1.7 Illustration of axial elongation and lateral contraction of a rod under uni-axial loading P (Alderliesten, 2011, 1-7.jpg. Own Work.)

As mentioned in the introduction, the diameter of a spring becomes smaller than its initial diameter when loaded. A similar contraction can be observed in any material under axial loading in the linear elastic part of the curve. This transverse contraction is illustrated in Figure 1.7.

To visualise this contraction in transverse direction, one may as a first qualitative illustration consider the elongation of rubber. During elongation, rubber will not only elongate but also become thinner. Elongation in lateral (loading) direction must then be compensated by transverse contraction.

Quantitatively, this visualisation is incorrect. The exact amount of contraction during uni-axial loading is determined by the material. The relation between the lateral and transverse strain is represented by another constant

$$\varepsilon_t = -\nu\varepsilon_l = \nu\frac{\sigma_l}{E} \tag{1.5}$$

This constant ν is called the Poisson's ratio. Both the Young's modulus E and the Poisson's ratio ν are considered material constants.

1.6 SHEAR STRESS

In the case of shear or torsion loading (see Figure 1.4), shear stresses occur in the material. The shear stress τ is defined in a similar way as the normal stress; the force is divided by the area A to which it is applied, see Figure 1.8.

$$\tau = \frac{F}{A} \tag{1.6}$$

The relation between the shear strain γ and the shear stress is characterized by a relation similar to equation (1.4)

$$\tau = G\gamma \tag{1.7}$$

where G is the shear modulus of elasticity and γ the shear strain (equal to $\tan\theta$, see Figure 1.8). For linear elastic materials, there is a relation between E and G, given by

$$G = \frac{E}{2(1+\nu)} \tag{1.8}$$

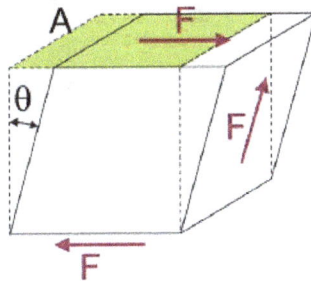

Figure 1.8 Illustration of the shear deformation due to shear forces acting on the surface of the element (TU Delft, n.d. 1-8.jpg. Own Work.)

1.7 BI-AXIAL LOADING

In the case of elastic bi-axial loading, i.e. loads are being applied in two directions, the stresses that occur in the material can be calculated using superposition.

This superposition is allowed, because the stress relates linearly to the load that is applied. If two load systems are applied simultaneously, the stress may thus be superimposed.

Figure 1.9 Illustration of a sheet loaded in either x-direction (TU Delft, n.d. 1-9.jpg. Own Work)

For the sheet loaded in x-direction (Figure 1.9), the strains in both directions can be given by

$$\varepsilon_x = \frac{\sigma_x}{E} \quad ; \quad \varepsilon_y = -\nu\frac{\sigma_x}{E}$$

(1.9)

For the sheet loaded in y-direction, the strains are given by

$$\varepsilon_x = -\nu\frac{\sigma_y}{E} \quad ; \quad \varepsilon_y = \frac{\sigma_y}{E}$$

(1.10)

Superimposing both load cases, as illustrated in Figure 1.10, the stresses in equation (1.9) and (1.10) can be superimposed. This gives

$$\varepsilon_x = \frac{\sigma_x}{E} - \nu\frac{\sigma_y}{E} \quad ; \quad \varepsilon_y = -\nu\frac{\sigma_x}{E} + \frac{\sigma_y}{E}$$

(1.11)

Equation (1.11) is known as *the Hooke's law* for a sheet in bi-axial stress condition.

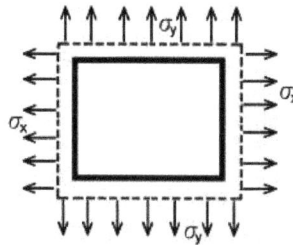

Figure 1.10 Illustration of a sheet loaded in lateral and transverse direction simultaneously (TU Delft, n.d. 1-10.jpg. Own Work.)

1.8 STIFFNESS AND APPARENT STIFFNESS

Stiffness, expressed by the Young's modulus, is the material's resistance against deformation, see Figure 1.5. A higher stiffness (E-modulus) means that a higher force must be applied to obtain a specified elongation (or strain).

For a uni-axially loaded sheet, the stiffness relates directly to the Young's modulus, E. In a bi-axially loaded situation however, the apparent stiffness may be different fromthe material stiffness. This can be illustrated with the Hooke's law, given by equation(1.11), and a wide sheet clamped at both sides over its full length and loaded in one direction, see Figure 1.11.

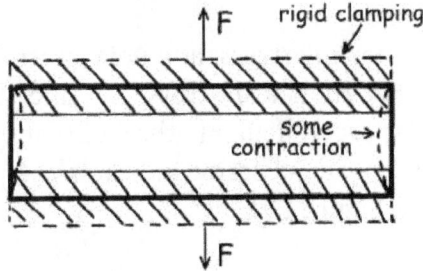

Figure 1.11 Illustration of a wide sheet rigidly clamped at both ends and loaded in lateral direction (TU Delft, n.d. 1-11.jpg. Own Work.)

The transverse strain ε_x is equal to zero as the clamping prohibits the contraction.With equation (1.11) this implies that

$$\varepsilon_x = 0 = -\nu\frac{\sigma_y}{E} + \frac{\sigma_x}{E} \quad \rightarrow \quad \sigma_x = \nu\sigma_y \tag{1.12}$$

Substitution of this relation between the longitudinal and transverse strain into the expression in longitudinal direction, equation (1.11), yields

$$\varepsilon_y = \frac{\sigma_y}{E} - \nu\frac{(\nu\sigma_y)}{E} = (1 - \nu^2)\frac{\sigma_y}{E} \tag{1.13}$$

The strain in a regular tensile test, where the transverse contraction is not prohibited ($\sigma_t=0$), is given by

$$\varepsilon_y = \frac{\sigma_y}{E} \tag{1.14}$$

This means that the apparent Young's modulus is given by

$$E^* = \frac{1}{1 - \nu^2}E \tag{1.15}$$

1.9 ISOTROPIC AND ANISOTROPIC SHEET DEFORMATION

An isotropic sheet is a sheet that is considered to have equal properties in any direction of the sheet. For the tensile (longitudinal and transverse) and the shear deformation of such a sheet, see Figures 1.13 and 1.14, the equations are obtained by combining equations (1.7) and (1.11)

$$\varepsilon_x = \frac{\sigma_x}{E} - \nu\frac{\sigma_y}{E}$$

$$\varepsilon_y = -\nu\frac{\sigma_x}{E} + \frac{\sigma_y}{E}$$

$$\gamma_{xy} = \frac{\tau_{xy}}{G}$$

(1.16)

The subscripts for the normal stress and normal strain indicate the direction of the stress and strain. For the shear stress   and shear strain γ_{xy} the first subscript indicates the axis perpendicular to the face that the shear stress and strain are acting on, while the second subscript indicates the positive direction of the shear stress and strain, see Figure 1.12.

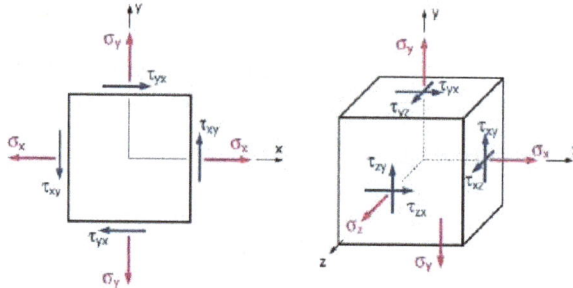

Figure 1.12 Illustration of normal and shear stresses acting on a two-dimensional and three-dimensional element (Alderliesten, 2011, 1-12.jpg. Own Work)

This set of equations can also be written in matrix formulation

$$\begin{bmatrix} \varepsilon_x \\ \varepsilon_y \\ \gamma_{xy} \end{bmatrix} = \begin{bmatrix} \frac{1}{E} & \frac{-\nu}{E} & 0 \\ \frac{-\nu}{E} & \frac{1}{E} & 0 \\ 0 & 0 & \frac{1}{G} \end{bmatrix} \begin{bmatrix} \sigma_x \\ \sigma_y \\ \tau_{xy} \end{bmatrix}$$

(1.17)

An anisotropic sheet has different properties in the different material directions. An example of an anisotropic sheet can be the fibre reinforced ply. Because the tensile and shear deformation is dependent on the properties in the particular directions, see Figure 1.13 and Figure 1.14, equation (1.17) must be extended to

$$\begin{bmatrix} \varepsilon_x \\ \varepsilon_y \\ \gamma_{xy} \end{bmatrix} = \begin{bmatrix} \frac{1}{E_x} & \frac{-\nu_{yx}}{E_y} & 0 \\ \frac{-\nu_{xy}}{E_x} & \frac{1}{E_y} & 0 \\ 0 & 0 & \frac{1}{G_{xy}} \end{bmatrix} \begin{bmatrix} \sigma_x \\ \sigma_y \\ \tau_{xy} \end{bmatrix}$$

(1.18)

The subscript 'xy' for the Poisson's ratio describes the contraction in y-direction for an extension (direction of load) in x-direction.

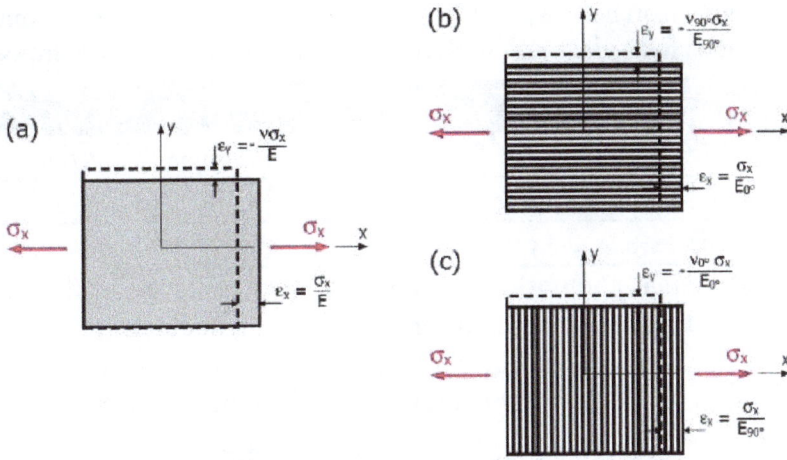

Figure 1.13 Illustration of tensile deformation under tensile stress for isotropic sheet (a), and anisotropic sheet (b,c) (Alderliesten, 2011. 1-13.jpg. Own Work.)

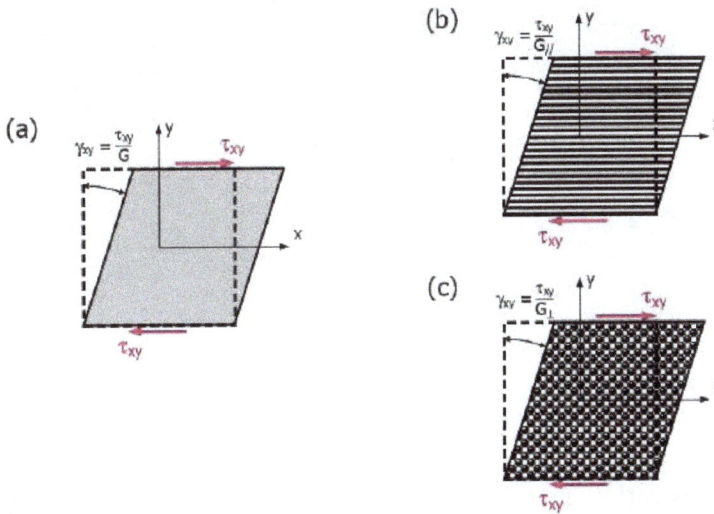

Figure 1.14 Illustration of shear deformation under shear stress for isotropic sheet (a), and anisotropic sheet (b,c) (Alderliesten, 2011. 1-14.jpg. Own Work.)

Here, one must realise that the excellent stiffness and strength properties often given for composite materials may be given for the longitudinal direction only. Table 1.1 illustrates the strength and stiffness properties for two thermoset fibre reinforced composites in the two principal material directions. Indeed, the stiffness and strength is significant in fibre directions, but perpendicular to the fibres the properties are verylow.

Table 1.1 Comparison between stiffness and strength in the two principal material directions forE-glass and high modulus carbon thermoset composite

Material	E_x	E_y	σU_X	σU_Y
	[kN/mm²]	[kN/mm²]	[N/mm²]	[N/mm²]
E-glass epoxy (Uni-Directional) UD-60%	45	8	1020	40
High modulus (HM) carbon epoxy UD-60%	220	10	760	40

As a consequence, composite (aeronautical) structures are made of alternating various plies that are oriented in different directions to obtain sufficient strength in each direction. The amount of fibres in each direction may vary depending on the loads cases. This design freedom is illustrated by the shaded area in Figure 1.15.

In this figure, several laminate lay-ups are indicated to explain the presentation of this figure. One of the laminate lay-ups indicated in Figure 1.15 is the quasi-isotropiclaminate. Quasi-isotropy can be defined as the approximation of isotropy by orienting plies in different directions.

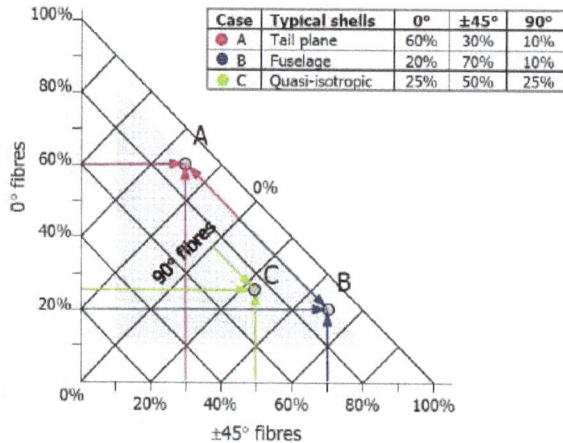

Figure 1.15 Illustration of the position of three typical laminate lay-ups in the design freedom for a fibre reinforced polymer composite panel mode of 0º, 90º, ±45º orientations only (Alderliesten, 2011. 1-15.jpg. Own Work.)

Table 1.2 Illustration of laminate properties for unidirectional E-glass epoxy plies (60%)

Orientation			Ex	Ey	σU_X	σU_Y	Note
0°	±45°	90°	[kN/mm²]	[kN/mm²]	[N/mm²]	[N/mm²]	
100%	0%	0%	45	8	1020	40	Unidirectional
0%	0%	100%	8	45	40	1020	Unidirectional
50%	0%	50%	~26	~26	~530	~530	Cross-ply
25%	50%	25%	~20	~20	~325	~325	Quasi-isotropic

The consequence of the combination of various orientations in a laminate lay-up is that the mechanical properties of the laminate are generally an average of the individual ply properties. This is illustrated in Table 1.2 with the example of E-glass epoxy from Table 1.1.

The stress-strain relationship for a composite material, i.e. a polymer (matrix) reinforced by fibres depends on the stress-strain behaviour of the individual constituents. For a single unidirectional ply this relationship is illustrated in Figure 1.16. From this figure, it can be understood that the stiffness of the ply is a function of the stiffness of the fibre and the matrix. The ply stiffness depends on the amount of fibres in the ply, which is described by the fibre volume fraction. For example, with 100% of fibres in the ply, the stiffness will equal the fibre stiffness, while with 0% of fibres the stiffness will equal the matrix stiffness. This linear relationship is called therule of mixtures and is discussed in chapter 3.

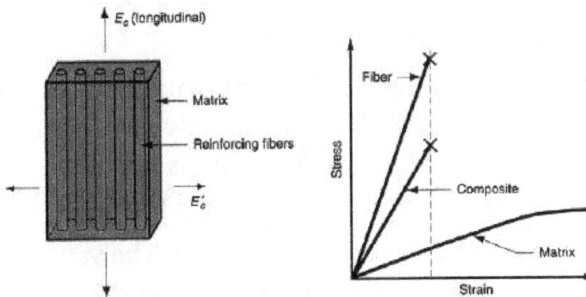

Figure 1.16 Illustration of the stress-strain curve for a fibre reinforced polymer (matrix) in relation to the constituent's stress-strain curve (TU Delft, n.d. 1-16.jpg. Own Work.)

Another observation from Figure 1.16, is that the strain to failure is in most cases not dependent on both constituents, but rather on the strain to failure of the fibres. Once the strain reaches the critical strain of the fibres, the fibres will fail, leaving the matrix with insufficient strength to carry the load, which will subsequently fail.

The earlier mentioned directionality of composite plies is important to consider. The high strength and stiffness of the composite may be described in fibre direction by the curve in Figure 1.16. However, perpendicular to the fibres, the strength and stiffness are described by the curve for the matrix, because there are no fibres in that direction to carry any load.

The directionality can be illustrated with the example shown in Figure 1.17. The high strength and stiffness of a composite may drop significantly to the low stiffness andstrength of the (unreinforced) polymer in the direction perpendicular to the fibres. This means that if sufficient strength and stiffness is required in different directions, multiple plies should be placed on top of each other, each oriented in a different direction. However, the consequence is that the strength of that lay-up is no longer equal to the single ply strength, but rather a function of the individual plies in their direction of loading. A first estimation of the laminate strength or stiffness can be made with again assuming a linear relationship (rule of mixtures).

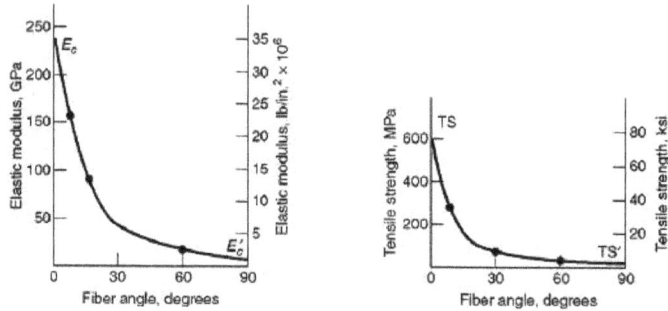

Figure 1.17 Relation between stiffness and strength of a composite ply and the angle or orientation of loading (TU Delft, n.d. 1-17.jpg. Own Work.)

Example: Laminate lay-up of multiple plies

Consider a laminate lay-up for vertical tail plane skins consisting of multiple plies for which the strength and stiffness of each ply are described by the curves in Figure 1.17. The lay-up is given by 60% of the fibres in 0°, 30% of the fibres in ±45° and 10% of the fibres in 90°. What is the stiffness of the laminate?

The modulus of elasticity is given in Figure 1.17. The values are approximately 240 GPa, 40 GPa, and 5 GPa for respectively 0°, ±45°, and 90°. The average stiffness of the laminate is proportional to the relative contribution of each ply.

This means that

$$E_{lam} = v_0 E_0 + v_{\pm45} E_{\pm45} + v_{90} E_{90}$$

$$= 0.6.240 + 0.3 . 40 + 0 . 1 . 5 = 156 . 5 \, G P a$$

where represents the laminate volume content of the plies in a given direction.

Since the vertical tail is primarily loaded in bending, most of the fibres are oriented in the span direction. However, for a fuselage a more quasi-isotropic lay-up is preferred because of the combined load cases in the fuselage. A typical lay-up that may be considered in that case is for example 20% of the fibres in 0°, 70% of the fibres in ±45° and 10% of the fibres in 90°. The laminate stiffnesswould then be

$$E_{lam} = v_0 E_0 + v_{\pm45} E_{\pm45} + v_{90} E_{90}$$

$$= 0.2.240 + 0.7 . 40 + 0 . 1 . 5 = 76 . 5 \, G P a$$

Note however, that this laminate has a stiffness in ±45° direction that is at least twice as high.

1.10 TOUGHNESS

The toughness of a material is often considered important in aeronautical structures because it represents the resistance of the material against fracture, formation of damage or impact. This parameter relates directly to the damage tolerance concept (see chapter 9) applied to ensure structural integrity during the entire operational life of, for example, an aircraft.

The toughness of a material is defined as resistance against fracture, and it is in general considered to be represented by the area underneath the stress-strain curve, see Figure 1.18. This area represents the mechanical deformation energy per unit volume prior to failure. Evaluating the units related to the area underneath the stress- strain curve, it can be shown that the unit of toughness is J/m³, which is the energy [J] per unit volume.

$$\sigma \cdot \varepsilon = \frac{F}{A} \cdot \frac{\Delta L}{L} = \left[\frac{N}{m^2}\right] \cdot \left[\frac{m}{m}\right] = \left[\frac{Nm}{m^3}\right] = \left[\frac{J}{m^3}\right]$$

(1.19)

Aside from the toughness, often different definitions are considered. For example, the impact toughness is the minimum energy required to fracture a material of specified dimensions under impact. This energy is not only dependent on the material itself, but also on the dimensions of the sample being fractured. Therefore, the test to determine the fracture toughness and the specimen dimensions are prescribed in testing standards to enable correlation of different materials. The set-up and specimen are illustrated in Figure 1.19.

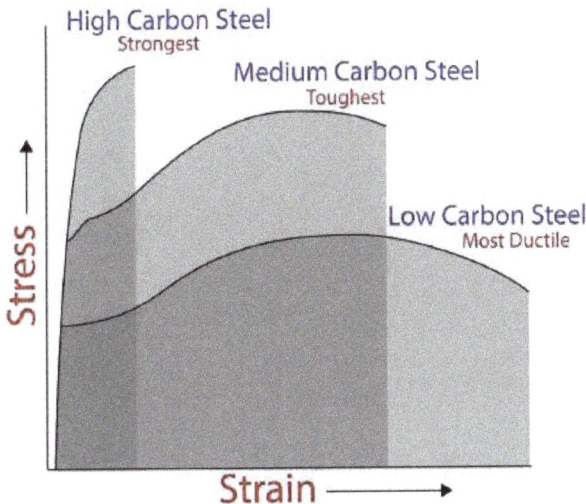

Figure 1.18 Three example stress-strain curves with the area underneath the curve shaded; the curve with the largest shaded area is considered to represent the toughest material. (TU Delft, n.d. 1-18.jpg. Own Work.)

Another important toughness parameter is the fracture toughness. This parameter represents the resistance of a material against fracture in presence of a crack. Thereis an important difference between toughness and fracture toughness. Although the area underneath the stress-strain curve, see Figure 1.18, qualitatively relates to the fracture toughness, the relation is not as straightforward as with toughness.

Materials with high fracture toughness usually fracture with significant ductile deformation, while materials with low fracture toughness fail in a brittle manner. In general, to fracture a material with high fracture toughness, a lot of energy or load is required, which implies that these materials are preferred for damage tolerant designs.

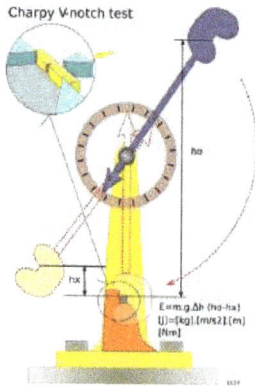

Figure 1.19 Test set-up for impact toughness measurements (left, Laurensvanlieshout, 2017, CC-BY-SA 4.0) and an intact and factured impact toughness specimen (right, Otrebski, 2013, CC-BY-SA 3.0)

CHAPTER-2

ENVIRONMENT AND DURABILITY

2.1 INTRODUCTION

The material properties discussed in the previous chapter are considered to describe the behaviour of a specific material. However, this does not mean that the properties are constant under all conditions. Most material properties change for example with temperature. Increasing or decreasing the temperature will affect material properties like stiffness and strength.

Another aspect that should be considered is the duration of operation. An aircraft being operated for instance for 30 years will face degradation of structural and mechanical behaviour due to environment effects.

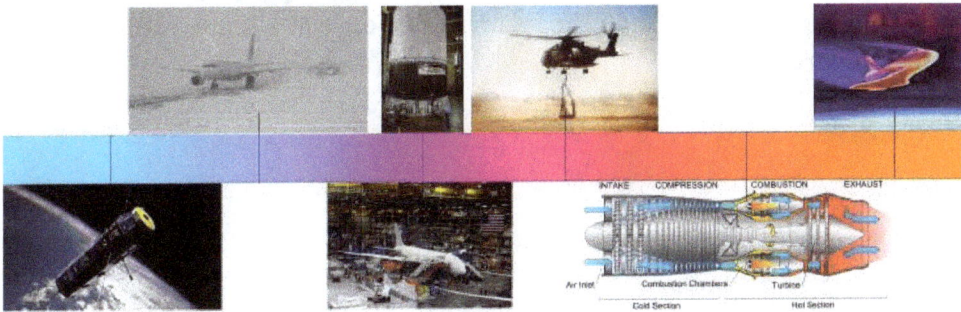

Figure 2.1: Illustration of structural aircraft and spacecraft applications and the temperature ranges in operation. Derivative from top left: Koul, (2008), CC-BY-NC 2.0, top middle: Deaton – NASA, (2013), Public Domain, top right: Stier, (2009) Open Government License, bottom left: NASA, (2006), Public Domain, bottom middle: NASA, (2011) Public Domain, bottom right: Jetstar Airways, (2013), CC-BY-SA 2.0.

The influence of the environment in which the structure or component will be operated is important to consider. For example, an engineer or designer should consider that if a structure is required to withstand certain loads during operation, the material strength may reduce for specific operating conditions. This may be either high temperatures of the environment, or degradation throughout the life of the structure due to aggressive environments. An illustration of typical applications and operationing (temperature) conditions is given in Figure 2.1.

This chapter tries to describe the effect of the environment on the material and structure, so as to increase awareness of this aspect to future engineers.

2.2 THE EFFECT OF AMBIENT TEMPERATURE

The effect of the ambient temperature on the material properties can be illustrated with the data given in handbooks as for example Metallic Materials Properties Development and Standardization handbook [1]. In Figure 2.2, an example is given for the effect of ambient temperature on the ultimate and yield strength of 2024-T3 aluminium sheets in a temperature range below the melting temperature of the alloy (T_m = 500-640 °C).

In this figure it is also demonstrated that the duration of exposure to that temperature may have a considerable effect above certain temperatures. The 2024-T3 aluminium alloy is widely applied in aeronautical structures. The nominal maximum operational temperature for this alloy is often specified to be about 135°C. From Figure 2.2 it is evident that above this temperature, the mechanical properties will drop rapidly, especially when exposed for longer times.

Metallic materials are not the only materials that show dependency of mechanical properties on the ambient temperature. In general, all engineering materials exhibit temperature dependent material behaviour. Especially in polymers one may also observe a transition in the material response at a temperature below the melting temperature.

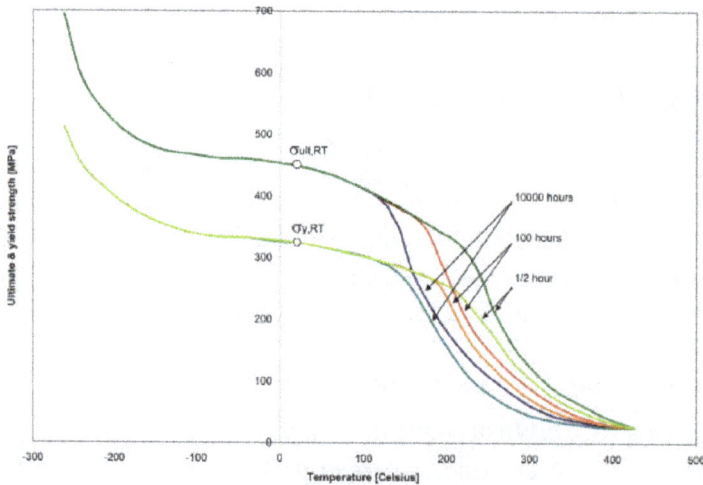

Figure 2.2: Effect of temperature on ultimate and yield strength for 2024-T3 sheets based on data from Rice et al., (2003). (Alderliesten, 2011. 2-2.jpg. Own Work.)

The transition in general relates to the transition from a solid state of the material into a rubbery state. The temperature at which this transition is observed is called the glass-transition temperature T_g. This refers to the transition glass exhibits at elevated temperatures, exploited in the glass blowing process. The phenomenon is illustrated in Figure 2.3 for the modulus of elasticity. However, the effect is also evident for the strength and strain to failure; increasing the temperature beyond the transition temperature decreases the strength of the material, while the strain to failure is oftenincreased.

For structural applications this implies that operational temperature may never approach the glass transition temperature, otherwise the mechanical properties would drop significantly risking premature failure of the structure.

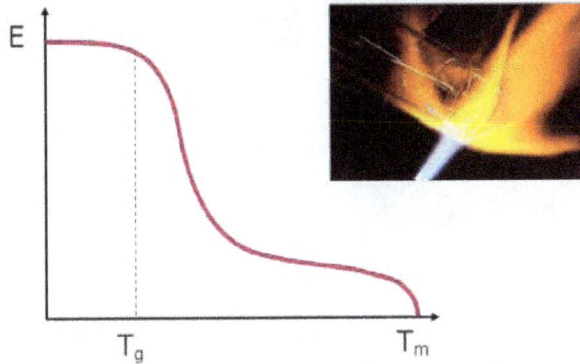

Figure 2.3: Illustration of the transition in modulus of elasticity near the glass transition temperature Tg. Derivative from left: Alderliesten, (2011), 2-3a.jpg, Own Work, and right: Anon., (2006), CC-0.

2.2.1 Effect of elevated temperature

In general, the effect of increasing the temperature is that most mechanical and fatigue properties of engineering materials deteriorate. This is also clearly illustrated in Figure 2.2. This means that it has to be verified that the mechanical properties of the selected materials remains above the specified levels within the full operational temperature range.

For the case of 2024-T3, this means that the specified ultimate and yield strength are minimum values that are lower than the values obtained at room temperature. To determine the minimum allowable strength of the material, knock-down factors are applied to the values obtained at room temperature.

For many metallic structures, the reduction of yield strength may not directly implicate a safety issue, because the ultimate strength may still be considerable. However, the application of stresses beyond the reduced yield strength may cause permanent (plastic) deformation.

Another aspect related to elevated temperature, especially high temperatures, is the creep phenomenon. Creep is a small, but steady ongoing deformation of materials under the application of constant stress. Although these stresses can be below the yield strength of the material, the ongoing deformation may still occur. At room temperature and low temperatures, this phenomenon is usually insignificant. At elevated temperatures, especially high temperatures near the melting temperatures, this phenomenon may cause permanent deformations within limited times of load application. The deformation rate is thus dependent on the applied load level, the temperature level, and the mechanical properties of the material.

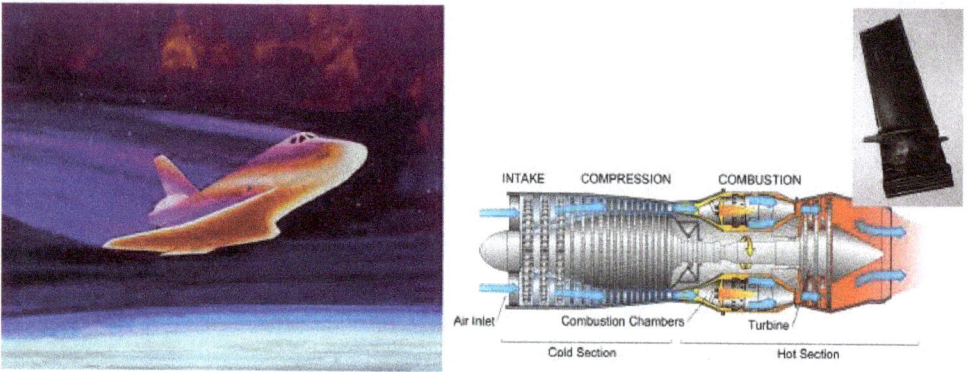

Figure 2.4: Illustration of high temperature applications; thermal protection systems and engine turbine blades. Derivative from left: NASA, (2006), Public Domain, middle: Dahl, (2007), CC-BY-SA 4.0 and right: Saunders-Smits, (2018), 2-4-c.jpg. Own Work.

For certain high temperature applications, see Figure 2.4, the creep phenomenon may significantly limit the amount of materials that can be applied. For example engine turbine blades are exposed for a long time (duration of a long distance flight) at high temperature, while constantly exposed to significant centrifugal loads. These components are therefore specifically designed against creep. For example, single crystal alloys (Ni-based alloys) are developed that have significant creep resistance.

2.2.2 Thermal stresses

Aside from the effect the environmental temperature has on the mechanical properties of the applied structural materials, the designer has to consider thermal stresses. The material will expand or contract with respectively increasing or decreasing temperatures. The relation between the temperature and the expansion is described by the volumetric thermal expansion coefficient.

$$\alpha = \frac{1}{V}\frac{dV}{dT}$$

(2.1)

Which is considered a material property. In equation (2.1) V is the volume and dV/ dT the expansions rate of the volume with the temperature. For isotropic materials, the coefficient of thermal expansion is identical in all principal material directions. However, for anisotropic materials, the coefficient is different for the different directions, like the other mechanical properties.

The different expansion coefficients for different materials, see Table 2.1, implies that a composite or hybrid structure, i.e. a structure comprising multiple materials, will face differences in expansion. Beside the mechanical loads and corresponding stresses that are excerted to the structure, these differences in thermal expansion may impose additional stresses once free expansion is prohibited.

Table 2.1 Linear coefficients of thermal expansion for different materials

Material	α_x	α_y
	[1/ºC]	[1/ºC]
Titanium Ti-6Al-4V (Grade 5)	9.2.10-6	9.2-10-6
Aluminium 2024-T3	22.10-6	22.10-6
Magnesium AZ31-H24	26.10-6	26.10-6
S2-glass epoxy UD-60	26.2-10-6	6.1-10-6
%Carbon epoxy UD-60%	-0.4-10-6	27.10-6

The significance of this aspect may be illustrated with press releases on the Boeing 787, where it was reported that the aluminium shear ties that fixate the fuselage frames to the composite skin in the rear fuselage section had to be replaced (Cohen 2010). The initial design did not account for the repeated cooling and warming of the unpressurized aft fuselage section 48. As result of these temperature cycles, the shear ties may repeatedly pull away from the skin with potential influence on the integrity of the structure.

Although this design flaw was detected prior to any 787 delivery, it emphasizes the importance of accounting for potential additional loading due to thermal stresses.

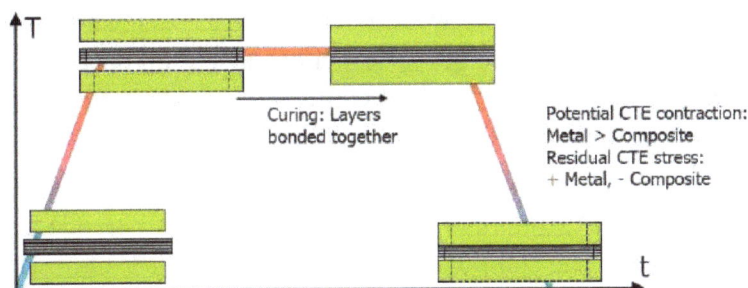

2.2.3 Effect of low temperatures

Obviously, the effect of thermally induced residual stresses is also present when ambient temperatures are decreased to low temperatures. For the Fibre Metal Laminate Glare for instance, the residual stresses increase further with decreasing temperatures. This is because the difference with the curing temperature, *i.e.* $\Delta T = T_{cure} - T$ is further increasing.

However, in these laminates the mechanical and fatigue properties are becoming better despite the increasing residual stresses. This is related to the general influence low temperatures have on metallic materials.

At lower temperatures, the micromechanical response of materials results in higher resistance against elastic and plastic deformation. A higher resistance against deformation relates to an increase in modulus of elasticity and yield strength of materials.In addition, the chemical reaction and diffusion rates decrease at lower temperatures. To some extent this is the consequence of the lower water vapour pressure. Because there is less water vapour in the air at low temperatures, the chemical reaction with materials reduces.

A special case of the effect of low temperatures is the transition in fracture and impact toughness observed in some low carbon steel alloys. This transition relates to the change in fracture appearance. Where at room temperature, the fracture is completely ductile (high toughness) the fracture changes to brittle at low temperatures (low toughness). Examples of this phenomenon are the failures of the Liberty ships and T2 tankers, shown in Figure 2.6.

2.3 THE EFFECT OF HUMIDITY

In general, a humid environment has a detrimental effect on the structural propertiesof both metallic and composite structures. However, the reason for the deterioration of both material types is different. Metallic materials in a humid environment may be more affected by corrosion attacks that damage the material and reduce the effective thickness of the structure or component. However, a composite structure in a humid environment faces ingress of moisture into the polymer matrix, which deteriorates both the cohesive strength of the polymer, but also the adhesive strength of the bond between fibre and matrix. Thus where in metallic materials the strength relates to reduction of cross-section because corrosion has eaten away the material,the strength of composites reduces due to the reduction in chemical bonding and softening of the matrix.

In both cases, time is an important parameter. The longer a structure is exposed to a humid environment, the more time there is to either corrode a structure, or for moisture to ingress the composite. In general, the reduction in strength due to environmental attacks and humidity is dependent on the exposure time.

Here an interesting difference can be observed between the performance of a fibre reinforced polymer composite and a Fibre Metal Laminate. Because the metallic sheets do not allow moisture to penetrate the material, the moisture ingress in FMLs is in general limited to edges of panels and cut-outs and drilled holes (for riveting for example). The problem then reduces to a 2-dimentional problem. A fibre reinforced polymer, or carbon fibre composite material is sensitive to moisture ingress from all sides, which implies a 3-dimensional problem. Where FMLs required additional protection (coating) for edges only, the composite structures require specific coating applied to the structure.

Example: T2 and Liberty ships

On 16 January 1943, 24 hours after being released from the shipyard, the T2 tanker S.S. Schenectady broke mid ships into 2 pieces in the docks near Portland, Oregon. This ship was the first ship of a new series built. Although hull fractures had occurred occasionally before, this failure occurred with a brand new ship while being in the docks.

The T2 tankers and Liberty ships where ships that were manufactured quickly, within about 5 days, to provide transport to the fleet at a higher rate than German submarines could destroy. Where in the years 1930-1937 about 71 merchants ships were built in the USA, 5777 ships were built between 1939 and 1945.

Figure 2.6: The T2 tanker S.S. Schenectady broke in 2 pieces on 16 January 1943, being 24 hours old (Derivative from US GPO, 1943, Public Domain)

Instead of riveting, welding was applied, which did not only increase the speed of production, but also enabled reduction of structural weight. Initially, the main reason was considered to be bad welding, but further investigation made clear that the steel used for construction appeared to be sensitive to low temperature. At certain temperature levels, the impact and fracture toughness exhibit a significant transition. The reduction in toughness relates to the transition from ductile fracture (evidence of high energy absorption) to brittle failure (low energy absorption prior to fracture). This phenomenon is illustrated in Figure 2.7.

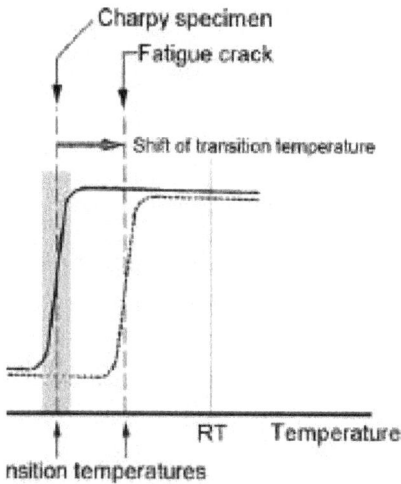

Figure 2.7: Illustration of the transition in impact energy (left) and the corresponding ductile and brittle fracture (right). Derivative from left: TU Delft, (n.d.), Own Work, and right: Broekhuis, (2018), Own Work.

2.4 ENVIRONMENTAL ASPECTS

The discussion in this chapter has been limited up until now to the effect of temperature and humidity. However, for both aeronautical and space structures various environments can be distinguished that each have their particular influence on the mechanical performance of a material or a structure.

The following typical environments may be identified:

- Air/moisture/salty environment
- Space and re-entry
- Fuel exposure
- Exposure to hydraulics
- Exposure to cleaning agents

2.4.1 Air, moisture and/or salty environment

Depending on the type of material, there can be different environments considered to be harmful or detrimental. In general, the detrimental processes due to aggressiveenvironments are accelerated with increasing temperatures.

For most metal alloys the combination of moisture and salty environment forms the aggressive environment that may lead to corrosion. Corrosion is the electrochemical reaction that metallic atoms have with oxidants in the environment. This rate at which this process may occur depends on the environment. In a humid environment or water, corrosion occurs generally faster than in (dry) air. The oxidant may be for instance oxygen.

Figure 2.8: Illustration of a humid and salty environment (left) and the consequence of corrosion for an aluminium structure (right). Derivative from left: Skeeze, (2008), CC0, and right: Saunders-Smits, (2017), Own Work.)

For fibre reinforced composites usually the combination of humidity with higher temperatures forms the environment that may lead to material degradation or loss of structural properties. The earlier mentioned glass transition temperature may reduce to lower temperatures under the influence of moisture, with reduction in strength and stiffness at lower temperatures.

2.4.2 Space and re-entry

Space structures operate under different conditions as for example aircraft structures. To begin with, the temperature range under which most space structuresoperate is significantly larger than for aircraft structures, see Figure 2.1.

But aside from the temperature aspect, the environment and the relevant aspects to be considered are substantially different as well. Where in aircraft structures, environmental degradation may be attributed to moisture and oxygen in combination with temperature effects, typical environments and environmental aspects considered for space structures are;

- Radiation/UV exposure
- Free radicals, atomic Oxygen (O+)
- Vacuum (outgassing)

Ultraviolet (UV) light is electromagnetic radiation with a wavelength ranging between10 nm and 400 nm (shorter than that of visible light). A lot of natural and synthetic polymers deteriorate under UV exposure. Here, fibres that are known to be sensitive to UV radiation are for example aramid fibres, like Kevlar.

Figure 2.9 Illustration of surface of the Space shuttle (left) and the consequence of the aggressive environment on the structure's surface (right). Derivative from left: Landis – NASA, (2009) Public Domain, and right: Tschida – NASA (2005) , Public Domain.

In the outer atmosphere, free radicals, especially atomic oxygen, play an important role in the degradation of materials and structures. The amount of atomic oxygen relates to the altitude and the activity of the sun.

Different structural materials respond differently to the exposure to atomic oxygen. Aluminium for example erodes slowly under atomic oxygen exposure, while gold and platinum are highly resistant. A lot of polymers are known to be very sensitive and require the application of special coatings (for example silicon based coatings) and paints to protect the structure from atomic oxygen erosion.

Especially in vacuum, outgassing is an important topic of concern. Many materials, like for example polymers, composites, adhesives, are based on dissolvers, or contain substances that can evaporate from the material. But even metals may release gasses from cracks or impurities in the material.

In general, the consequence of degassing on the material or structure is that the mechanical properties of the material may deteriorate in time. Also the released gasses may condense on other cold surfaces causing trouble to the operation of certain components, for instance solar cells and telescope lenses.

Also here the temperature may have an acceleration effect; at higher temperatures, the chemical reaction rate with the material and the vapour pressure increase.

2.4.63 Exposure to fuel or hydraulics and cleaning agents

Other than air, moisture for aeronautical structures and the environments discussed in the previous section, environments should be considered that may cause degradation of structure and corrosion of materials. Especially for aircraft structures, several additional environmental aspects should be considered, of which some as listed below;

⊙ Fuel

- ⊙ Hydraulics
- ⊙ Cleaning agents

In case of an integral fuel tank, see Figure 2.10, the fuel is kept inside the structure without use of additional fuel bags. All joints and structural connections are sealed air and liquid tight, to avoid leakage of the fuel.

This implies for that particular part of the structure, that the structural material is directly exposed to fuel. In order to avoid any degradation due to the fuel environment, one should then consider use of materials or coatings resistant to this type of environment.

This is especially an aspect to consider for polymers and fibre reinforced composites. Here, the question will be whether the polymers applied in the structural material contains dissolvers that may react with the chemicals in the fuel.

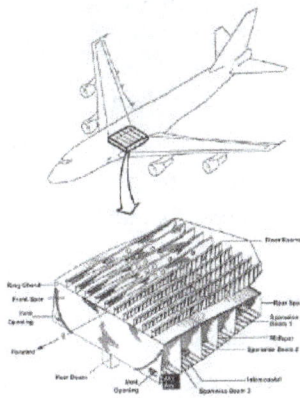

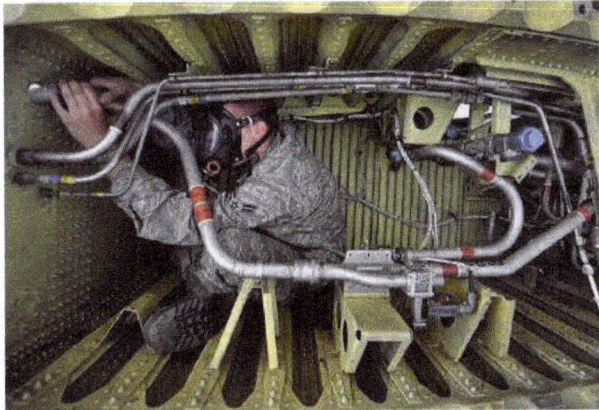

Figure 2.10 Illustration of integral fuel tanks, where the structure is directly exposed to kerosene fuel (left: NTSB, 2010, Public Domain; right: Britton – U.S. Airforce, 2010, Public Domain)

Another environmental aspect often not considered is related to operation and maintenance of structures. Often the selection of structural materials is thought of well, considering the environments and circumstances the structure will operate in with high probability.

However, for maintenance and operation one has to consider the materials applied in the structure. Use of cleaning agents to clean a dirty structure may impose structural degradation and impair the integrity of the structure, if the cleaning agents are basedon a chemical composition that reacts with the structural material. One has to specify the cleaning agents or at least the chemical basis of such agents for particular structure. Here, the example given in Figure 2.11 (right) illustrates that the specification works both ways. It reads: "Do not use on glass or aluminium." Manufacturers of cleaning agents define the restrictions to application of such cleaning agents.

In general, not only corrosion related to air and moisture, see section 2.4.1, should be considered in structural design, but also the deterioration due to other chemical environments. Here, one should consider that the general advantage claimed for carbon fibre composites is that this structural material does not corrode. However, depending on the epoxy system applied, it may deteriorate due to other environmental conditions, like for example UV exposure.

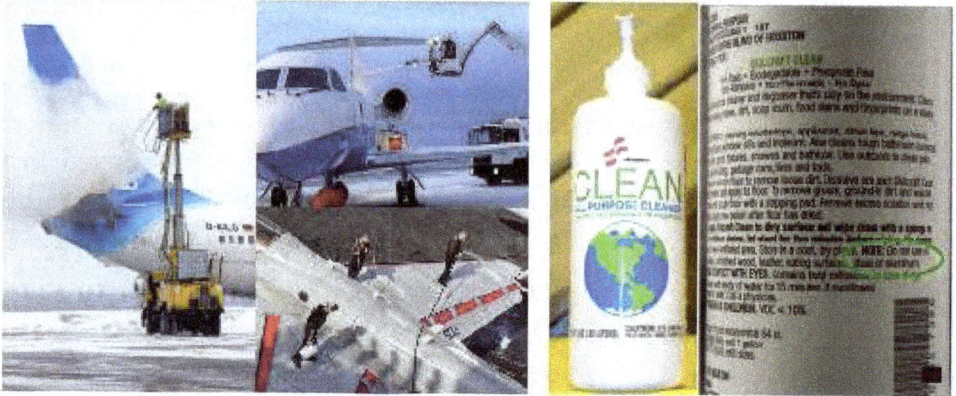

Figure 2.11: Photos of de-icing and cleaning procedures (extreme left: Brygg, 2009, CC-BY-2.0, middle left:Wollman – U.S.Navy, 2010, Public Domain) and an example of cleaning agent inappropriate for aluminium. (right, anon, n.d. Public Domain)

One type of corrosion should be added to the discussion here; galvanic corrosion. Galvanic corrosion is the electrochemical reaction process in which one metal may corrode due to electrical contact with another material or metal, while being in an environment that contains an electrolyte. This corrosion process forms the basis of batteries, where one metal corrodes to provide electrical current.

Especially in moisture rich environments, such contact between two materials may be easily made, which can cause corrosion of one of the metals involved. One example is given in Figure 2.13, where the aluminium rim corrodes in a wet, humid and potentially salty environment, due to the electrical connection with the chromium plated brass spoke.

It should be emphasized here however, that the contact does not necessarily be between two metals, it may also be between a metal and another material, of which the electric potential provides sufficient difference with the potential of the metal. For example, aluminium connected to carbon fibre reinforced composites, may lead to galvanic corrosion of the aluminium, due to the potential difference of these two structural materials. See for instance Figure 2.12, where the brass nipple of the spoke reacts with the aluminium rim.

The method to counteract galvanic corrosion is to isolate the different materials, avoiding the electrical connectivity, or to assure that the materials are not immersed in a solution containing an electrolyte.

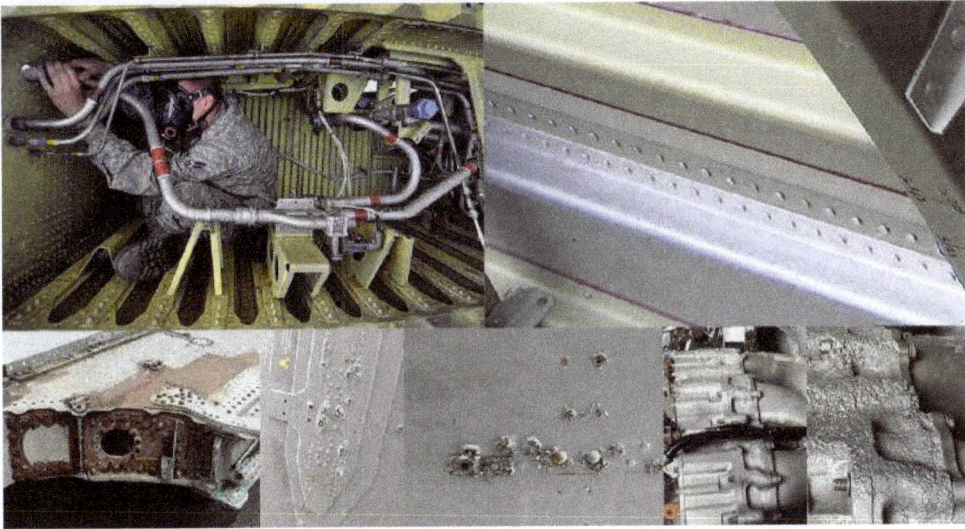

Figure 2.12: Examples of brand new virgin structures (upper row, left:Britton-U.S.Airforce, 2010, Public Domain, right TU Delft, n.d. Own Work.) and structures after decades of operational use (lower row: Saunders-Smits, 2017. Own Work.)

Figure 2.13: Example of galvanic corrosion on a bike in a corrosive (wet/humid/salty) environment (Hans, 2012, CC0); galvanic corrosion between chromium plated brass spoke nipple and aluminium rim (Open University, 2004, Copyright Open University)

Sometimes, the process is exploited as solution to counteract corrosion. Here, the example of placing zinc sacrifice material to steel (marine) structures could be mentioned. Since the zinc is less noble than steel, it will corrode first under a corrosive attack, protecting thereby the steel structure.

CHAPTER-3
MATERIAL TYPES

3.1 INTRODUCTION

The key difference between the structures and materials discipline and other disciplines related to flight is that this discipline is about materialisation of concepts also developed within the other disciplines. To create an aircraft or spacecraft one must use materials. Materials in that sense can thus be defined as substances, matters, constituents or elements that are used to build parts, components and structures.

The properties of materials do not depend on their geometry, but on their composition only. The relation between the composition and the properties of a material can be further explored, but for the time being, one may consider the properties as an artefact of materials.

There is a wide variety of materials available to be used in materialisation of components and structures. Typical examples of materials are metals (steel, aluminium, magnesium, etc), wood, ceramics, and polymers. All these materials have properties which do not depend on their shape, like for example mechanical properties, electrical properties, physical properties, etc.

However, to materialise an aircraft or spacecraft structure, certain material properties are required. As a consequence, not all materials available in this world can be used, or are preferred to be used. Aerospace structures require materials that are solids with good mechanical properties but with a low density. This class of materials is often referred to as lightweight materials. Since there are numerous materials that are lighter than the materials currently used in aerospace structures, a more appropriateindication would be lightweight structural materials.

The performance of materials should be as high as possible for the lowest possible weight. This can be phrased alternatively by stating that the performance to weight ratio should be as high as possible. This leads to the use of specific mechanical properties, which are the properties divided by the density or weight of the material. The use of those specific properties will be further discussed in chapter 8.

For application in aerospace structures, one can distinct the following material categories:

- ◉ Metal alloys
- ◉ Polymers
- ◉ Composites
- ◉ Ceramics

These categories are briefly discussed in the following sections. But before discussing these categories individually, one has to be aware that these materials have been retrieved from resources like ores (metal) and oil (composites and polymers). Once retrieved, they are transformed into semi-finished products like sheets, plates, bars, fibres, powder (polymers), etc. The semi-finished products are further processed into structural elements. For this transformation a huge number of processes are available that can be grouped into: casting, forming, machining, and joining processes. Subsequently, the structural elements are assembled into structures.

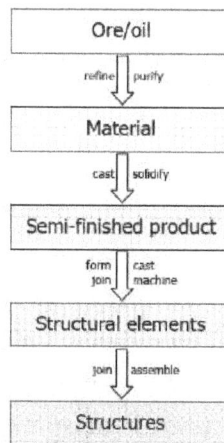

Ore/oil

refine | purify
⬇

Material

cast | solidify
⬇

Semi-finished product

form | cast
join | machine
⬇

Structural elements

join | assemble
⬇

Structures

Figure 3.1: Illustration of subsequent production steps from raw material resources to structures (Alderliesten, 2011. Own Work.)

The properties of structures are directly related to the material properties although they are not identical: structural properties are often influenced by the shape and geometry (design) too. However, there is also another aspect to be considered when optimizing between material and structural shape; not every structure or shape can be made of any material. Consider for example the Eiffel tower, the Parthenon, or a surf board. The selected materials (resp. metal, marble and composites) and the shapes of these artefacts are compatible.

This also implies that if the shape is not adapted to or compatible with the material, the material properties are not optimally used and exploited.

A similar relationship exists between material and manufacturing process. Metals can be melted, so casting and welding are available production processes for metals. These production processes cannot be applied to ceramics or fibre reinforced composites for instance.

The last relationship to mention is the one between the shape (or structure) and the manufacturing process. To fabricate a sheet metal wing rib, one may use a forming process. Replacing the same rib by a machined rib will consequently result in different details of the wing shape (local radii, thickness, etc). To put it the other way around: To create a cylindrical shape and a double-curved shape, different manufacturing processes are needed.

In summary: there is a strong interrelationship between the three entities "material", "structure or shape" and "manufacturing process". Changing one entity often affectsboth others. For the best solutions to structural problems, i.e. to truly optimize the structure and its performance, one should include all three aspects in the design andits evaluation. This is illustrated in Figure 3.2 .

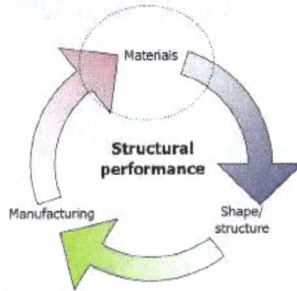

Figure 3.2: Illustration of the relation between Materials, Manufacturing and Design, with the topic of interest in this chapter highlighted (Alderliesten, 2011. Own Work.)

3.2 METAL ALLOYS

An alloy is made by adding alloying elements to the purified metal in order to increase or modify the properties of the pure metal. For example, adding a few percent of copper and magnesium to aluminium (like in Al-2024) increases the yield strength and ultimate strength both with a factor of 4 to 6. In general, metal alloys have good processibility, show plastic behaviour, and are rather cheap.

3.2.1 Typical mechanical properties

Metal alloys typically are isotropic materials exhibiting similar elastic properties in all directions of the material. Because of this isotropic behaviour, the material specifications and the specifications of physical and mechanical properties are often given indifferent of the orientation. Only for specific metals that show anisotropic behaviour, like for example aluminium-lithium alloys, and for rolled sheet material sometimes properties are specified in two directions. The orientation dependency for the rolled products is related to the shape of the grains (severely elongated in rolling direction) as a result of the rolling process.

Because metal alloys are ductile materials that yield beyond the yield strength, both ultimate and yield strength are specified. This value indicates how far the material can be loaded elastically before permanent plastic deformation may occur.

Table 3.1 gives some mechanical and physical properties of typical steel, aluminium, titanium and magnesium alloys. It can be observed from the data in this table that there is some relation between strength and stiffness on the one hand and the density of the material on the other hand.

Steel exhibits high strength and stiffness, but at the cost of a high density, whereas magnesium (the lightest alloy in the table) shows the lowest mechanical properties.

Table 3.1 Typical mechanical properties some metals

Metal	Alloy	E [GPa]	G [GPa]	σ_y [MPa]	σ_{ult} [MPa]	ε_{ult} [%]	v [-]	ρ [g/cm3]
Steel	AISI 301	193	71	965	1275	40	0.3	8.00
	AISI 4340	205	80	470	745	22	0.29	7.85
	D6AC	210	84	1724	1931	7	0.32	7.87
Aluminium	AA 2024-T3	72	27	345	483	18	0.33	2.78
	AA 7475-T761	70	27	448	517	12	0.33	2.81
Titanium	Ti6Al-4V (5)	114	44	880	950	14	0.34	4.43
Magnesium	AZ31B-H24	45	17	221	290	15	0.35	1.78

3.2.2 Typical applications

Typical applications for metals are structures and components that require high strength both in tension and in compression, see the examples in Figure 3.3. Example applications for steel alloys are found in aircraft (landing gear components), train components and rails, bridges, towers and cranes.

Aluminium alloys are for instance applied in the main fuselage and wing structure of most aircraft, train structures, and car and engine components.

In aeronautical structures, titanium is applied in applications that require performance at elevated temperatures, like for example in the Concorde and military fighters. Most magnesium alloys are not applied in aircraft for flammability risks.

In general, metal alloys are applied in components and products that are produced inhigh volumes. Examples here are the cars and cans.

Steel is also often applied as reinforcement material in for example civil applications. The application of steel cables in suspension bridges is an evident example. But alsoconcrete is reinforced with steel cables to increase the strength of the structure. Especially in case of high buildings the steel reinforcement is applied to pre-stress the structure, i.e. the steel reinforcement is put in tension (because of the excellent tensile properties), which by equilibrium puts the concrete in compression (for whichconcrete is known to perform excellent).

Figure 3.3: Typical applications of metals. Derivative from: Top left NI-CO-LE, (2017), CC0; Top right: Pingstone, (2004), Public Domain; Bottom left: KarinKarin, (2015), Public Domain; Bottom right: Bender, (2014), CC-BY-SA3.0.

3.3 POLYMERS

In general, polymers are not considered for structural applications. The polymers have relative low strength and stiffness and can therefore not be used as structural material. However, they are applied as structural adhesives to join other materials, and they are applied with additional reinforcement in composites.

3.3.1 Typical mechanical properties

Compared to rigid materials like metals, polymers exhibit significant lower stiffness and strength. Here, it should be noted that the stiffness of many polymers is not constant during loading. Whereas metals exhibit linear stress-strain behaviour during elastic deformation of the material, as illustrated in Figure 1.5, the stiffness of polymers often change with the amount of strain, see Figure 3.4. In case of such non-linear behaviour, the initial slope of the material is taken to determine the elasticmodulus.

Although the strength and stiffness are generally very low, the elongation at failure can be quite high. Some rubbers, for example, may strain up to 500% before failure occurs.

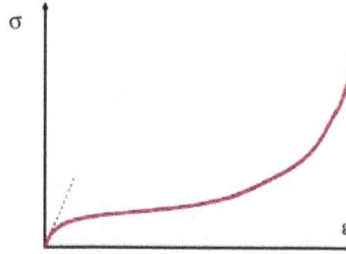

Figure 3.4: Qualitative illustration of the non-linear stress-strain behaviour of polymers (Alderliesten, 2011)

In chapter 2, it has been explained that the temperature has an influence on the mechanical properties of materials. Although this is in general the case for all materials, it is quite significant for polymers. Depending on the temperature, materials may either behave like brittle materials or like elastic materials. Especially at low temperature, many polymers show brittle behaviour.

With increasing the temperature a gradual transition can be observed from brittle to elastic and rubbery behaviour, while further increasing to high temperatures the material may become viscous or even liquid like.

This transition to the viscous state is important for polymers, because it implies a significant reduction in the mechanical properties. A well known transition for polymers is the so-called glass transition temperature.

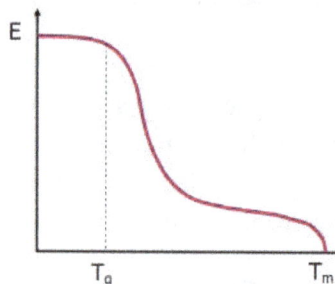

Figure 3.5: Transition in modulus of elasticity at the glass transition temperature, below the melting temperature (Alderliesten, 2011)

Some polymers exhibit different mechanical behaviour, depending on the rate they are strained. Glass fibres for example, exhibit higher strengths when loaded at very high rates. This can be beneficial in case of impact for example.

3.3.2 Typical applications

Because of the wide variety of polymers that exist, the number of applications is numerous. Some main categories can be distinguished:

- ◉ Elastomers
- ◉ Plastics
- ◉ Fibres

Some well known examples of polymer applications are illustrated in Figure 3.6. Rubbers are elastomers that are typically applied in tires, sealing, coatings and liners. They are in general characterized by their flexibility and the large strain to failures. Plastics can be divided into two main categories:

- ◉ Thermoplastic
- ◉ Thermoset

Figure 3.6: Typical applications of polymers. Derivative from Top left: Saunders-Smits (2018), 3-6-b.jpg. Own Work.; Top right: Yogipurnama, (2017), CC0; Bottom left: Anon., (2017), CC0; Bottom middle: Pexels, (2016), CC0; Bottom right: Hans, (2013), CC0.

Thermoplastic polymers melt when heated to certain temperatures and return to their glassy state when cooled again. These materials are often associated with weak Van der Waals forces. This means that the material can be melted above their melting temperature and moulded into components. The process is reversible, as reheating will melt the material again.

Thermoset materials however, are cured irreversibly, which means that once the chains link during curing the process cannot be reversed. These materials usually donot melt at high temperatures, but may decompose or burn when heated too high. The difference between these two materials is considered important, especially when addressing recyclability of the materials. Thermoplastic materials can be recycled relatively easy by heating above the melting temperature, while thermoset materials are in general not easy to recycle.

Example applications of thermoset materials are the old bakelite telephones and the epoxies used in fibre reinforced composites. Here, it should be mentioned that current developments seem to aim to replace, for certain composite applications, the thermoset matrix material by thermoplastic matrices.

Examples of fibre types are natural fibres, synthetic fibres and nylon. Application of these types of fibres in a fibre reinforced composite, implies that different polymers are combined into a structural material. The fibre is made of another polymer than the matrix material.

3.4 CERAMIC MATERIALS

Ceramics are not suitable for structures. They are too brittle and have poor processing features. However, they are applied in some space applications, for instance for thermal protection of the metallic or composite structure. Ceramics often consist of (metal) oxides and metals, in which ionic bonds between the different atoms providethe material structure.

3.4.1 Typical mechanical properties

In general, ceramics are hard and brittle materials that have very limited toughness due to the lack of ductility (small failure strain). In certain cases a high strength and stiffness can be achieved, but that depends on the composition of the material and the level of porosity. The reason why certain ceramics are considered for heat protection is that they are capable to sustain very high temperatures. Even at those temperature levels the bonds between the atoms remain very strong. This strong bond also implies that ceramics are often very resistant to wear.

3.4.2 Typical applications

A variety of typical applications for ceramics can be mentioned here. To start with the glass application, glass is applied in window panes, lenses, but also in fibres. Glass fibres are very stable fibres that have high mechanical properties both in tension andcompression. At high strain rate levels, the glass often provides a higher strength than when quasi-statically loaded to failure.

Another example of ceramics is clay. Porcelain and bricks are well known examples of these ceramics. In civil applications not only bricks, but also cement and lime are being applied as ceramic applications.

Other examples are cutting tools and abrasive materials due to its high wear resistance, armour reinforcement because of its high puncture resistance, and in case of glass fibres, due to its high impact resistance. The previously mentioned high heat resistance (1600 – 1700 °C) results in many applications in engine components and heat protection systems for, for example, the Space Shuttle. A selection of applications are illustrated in Figure 3.9.

Figure 3.9: Typical applications of ceramics. Derivative from Extreme left: Cjp24, (2007), CC-BY-SA3.0; Left: Torr68, (2005); CC-BY-SA3.0; Right: Anon.(n.d.), Public Domain; Extreme right: Atkeison, (2003) CC-BY-SA2.0.

Example: Space Shuttle Columbia

A known application of ceramic materials is the thermal protection tiles on the Space Shuttle. The importance of this protection is illustrated with the tragic accident on February 1, 2003. During its launch a piece of foam became detached from the tank and hit the leading edge of the wing causing damage to the ceramic skin. Although during lift-off and mission no apparent problems were observed, the Shuttle disintegrated during re-entry. Analysis revealed that during re-entry hot gasses could enter the wing structure through the damage affecting the structure behind the ceramic tiles.

In this accident the crew of 7 people were all killed.

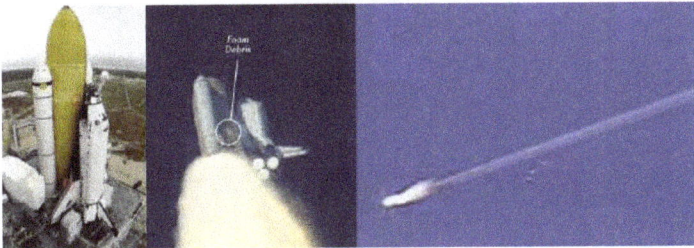

Figure 3.7: Space Shuttle Columbia (left) with indicated location of space debris (centre) and an image of the accident. Derivative from NASA, (2003), Public Domain.

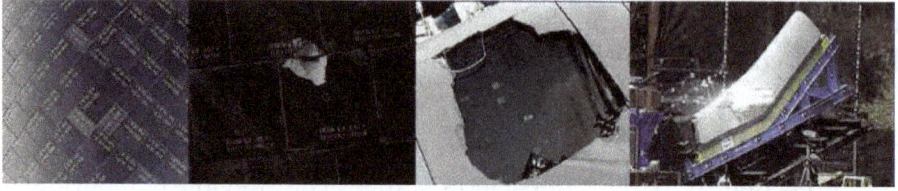

Figure 3.8: Photos of the ceramic tiles shown intact, with damage, test panel with damage and test set up. Derivative from Volk, (2008), CC-BY-SA2.0, and NASA, (2003 2007), Public Domain.

3.5 COMPOSITE MATERIALS

Composite materials are, as the name already indicates, materials that are composed of different materials. A more accurate description or definition is given by:

Composites are engineering materials, in which two or more distinct and structurally complementary substances with different physical or chemical properties are combined, to produce structural or functional properties not present in any individualcomponent.

An example of a composite is the fibre reinforced polymer composite, which consists of two distinct and complement materials, namely fibres and polymer. The function of the fibres is to reinforce the polymer providing strength and stiffness to the material and, by doing so, to carry the main portion of load. The function of the polymer is to support the fibres and to transfer the load to and from the fibres in shear. This is indicated in Figure 3.10.

Products and components made of fibre composites are fabricated with specific processes like filament winding, lay-up and curing, and press forming, discussed in the next chapter.

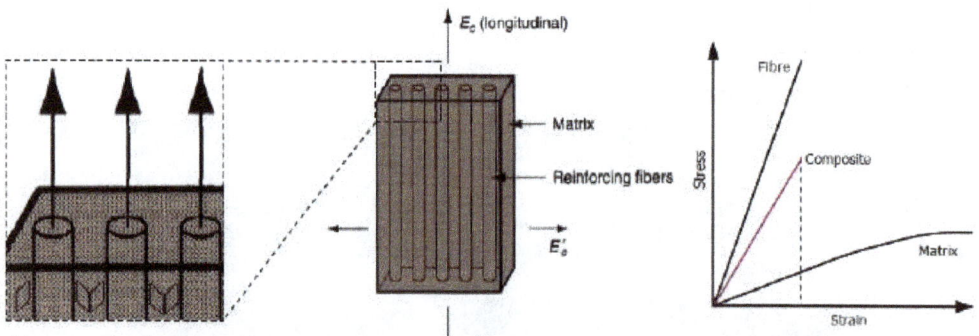

Figure 3.10: Illustration of a fibre reinforced polymer composite ply, and the related stress-strain behaviour of constituents and lamina (Alderliesten, 2011)

3.5.1 Typical mechanical properties

As is evident from Figure 3.10, the stress-strain behaviour of the fibre reinforced polymer composite is determined by the constituents of which it is composed. The stiffness of the lamina is a function of the stiffness of the polymer and the fibre, which can be estimated by the rule of mixtures, discussed in section 3.6. However, whereas the stiffness may be directly related to stiffness and volume content of each constituent in the lamina, the strain to failure is solely determined by the strain to failure of the fibres. Once the fibres fail, the strength of the remaining polymer is too low to carry the load.

One should pay attention to the definition given here for 'composites', because this definition states that any type of engineering structural material that satisfies this definition is considered to be a 'composite'. These days, people use the wording 'composites' often to indicate only one specific type of composites, namely the one constituted of carbon fibres and polymer. However, one should be aware that this is an inaccurate use of the definition of composites.

To illustrate the meaning of the definition of composites, another example of a composite is given by the category of hybrid materials, such as for example Fibre Metal Laminates (Vlot and Gunnink, 2001), see Figure (3.11). These structural materials consist of alternating metal and composite layers combining the benefit ofeach constituent material, while compensating each other's disadvantages.

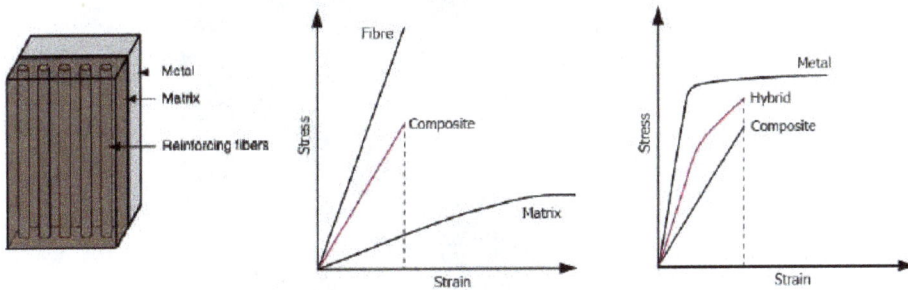

Figure 3.11: Illustration of a fibre metal reinforced polymer composite ply, and the related stress-strain behaviour of constituents and lamina (Alderliesten, 2011)

In general, fibre reinforced polymers are characterized by their high specific properties. The strength and stiffness to weight ratio is considerable. However, most composites behave elastic until failure, without showing any ductile behaviour. Despite the often very high strength and stiffness, this limits the toughness of these materials.

Due to the high directionality (fibre orientation) these materials enable tailoring to specific load applications (beams, cables, columns), but require multiple orientationsto cope with bi-axial load applications.

Table 3.2 Qualitative comparison of typical properties of several composites

Material	Specific strength	Failure strength	Electrical conductivity	Flame resistance	UV resistance	Chemical resistance
Glass fibre reinforced composite	High	Medium	Low	High	Medium	Low
Carbon fibre reinforced composite	High	Low	High	High	Medium	Low
Aramid fibre reinforced composite	High	Medium	Low	High	Low	Low
Fibre Metal Laminate	High	Medium	High	High	High	Medium

3.5.2 Typical applications

Typical applications are illustrated in Figure 3.12. Wind turbine blades are commonly made of glass fibre composites. Other applications are sail planes and pressure tanks and vessels.

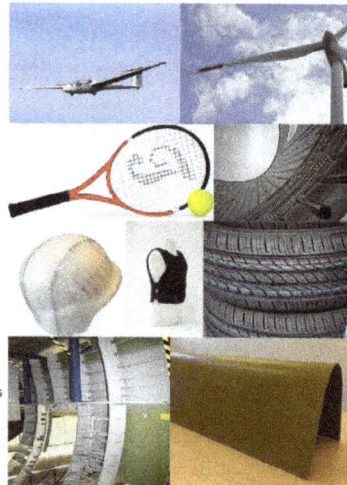

Glass fibre composites
- Wind turbine blades
- Sail planes
- Pressure tanks & vessels

Carbon fibre composites
- Automotive components
- Aerospace components
- Sailboats
- (motor) bikes
- Sport equipment

Aramid/kevlar composites
- Armor & bullet proof products
- Impact and penetration resistant products

Fibre Metal Laminates
- Upper fuselage skin panels
- Impact resistant leading edges
- Critical joint straps
- Lower wing panels

Figure 3.12: Typical applications of composites. Derivative from Top row, left: GuentherDillingen, (2012), CC0; Top row, right: medienluemmel, (2016), CC0; Second row, left: Gnokii, (2011), CC0; Second row, right: Boffoli, (2018), CC0; Third row left: Saunders-Smits, (2018), 3-12-f.jpg. Own Work.; Third row middle: PMulhalla, (2015), CC-BY-SA3.0; Third row right: Vinayr16, (2014), CC0; Bottow row: Saunders-Smits, (2018).

Carbon fibre composites are often applied in automotive and aerospace structures for their high stiffness. A well known application in sailboats is for example the mast.But also (motor) bikes are made of carbon fibre composites since the stiffness and rigidity of the frame is important in such design. Similarly certain sport equipment is made of these materials. Aramid and Kevlar based composites often find applications in armour and bullet proof protection systems, like bullet proof vests and cockpit doors that should resist terrorists. Also heat and flame resistant products are often made from aramid fibre reinforced composites.

Typical applications of the composite Fibre Metal Laminate (FML) concept are primarily found in aerospace applications. The reason is that these materials are specifically developed for their high strength and fracture toughness, which increases the damage tolerance of primary fuselage and wing structures, necessary for maintaining structural integrity. The FML Glare is currently applied as upper fuselageskin material and impact resistant empennage leading edges on the Airbus A380. The material is also applied as high damage tolerant butt strap joint material in the Airbus A340 fuselage.

3.6 RULE OF MIXTURES

A simple method to estimate the composite ply properties of a composite material isthe so-called rule of mixtures. This rule is a mean to *estimate* the lamina properties based on the properties of the individual constituents, i.e. fibre and matrix system. However, one should be aware that the method by no means is considered accurate.

$$M_{FRP} = M_F + M_M \rightarrow \rho_{FRP} V_{FRP} = \rho_F V_F + \rho_M V_M \tag{3.1}$$

where M indicates the mass of the constituent, V the volume and ρ the density. This equation can be written as

$$\rho_{FRP} = \rho_F \frac{V_F}{V_{FRP}} + \rho_M \frac{V_M}{V_{FRP}} \rightarrow \rho_{FRP} = \rho_F v_F + \rho_M v_M \tag{3.2}$$

where v indicates the volume fraction of the constituent in the fibre reinforced laminate. This linear relationship is illustrated for the density of the laminate in Figure 3.13.

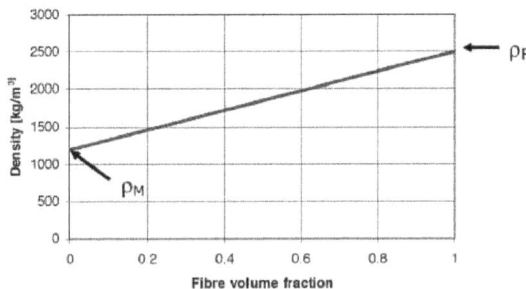

Figure 3.13: Rules of mixtures to estimate the composite ply properties based on the matrix and fibre properties relative to their volume content (Alderliesten, 2011)

Similarly, this rule of mixtures relationship is illustrated in Figure 3.14 for a carbon fibre composite with various lay-up configurations. Here, it should be clear that the high fibre volume may improve the properties, but that the different orientations reduce the overall laminate properties significantly. The grey shaded area in this figure illustrates the common fibre volume fractions typically applied in composites.

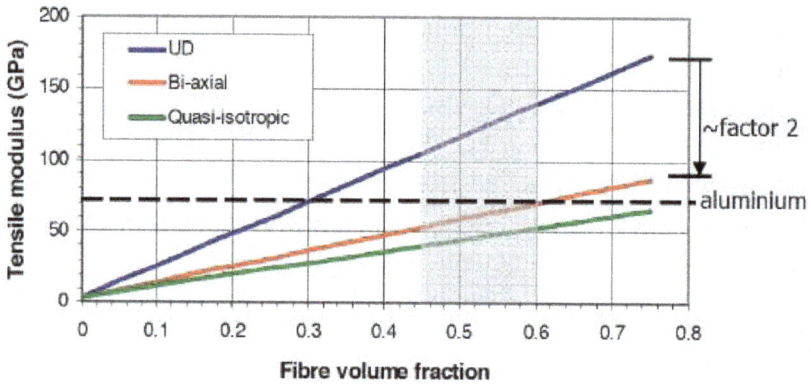

Figure 3.14: Illustration of the effect of fibre volume fraction of the individual composite plies and lay-up on the overall laminate stiffness (shaded area is typical range of fibre volume fractions). Alderliesten, (2011).

3.7 REQUIREMENTS FOR STRUCTURAL MATERIALS

One could assemble a list of requirements for the engineering materials considered. Comparison between these material requirements and, for example, structural requirements would reveal a large overlap. However, here one should be careful: there are significant differences between these two.

Several requirements for structures are also mentioned for materials: high strength, high stiffness, low weight, durability, and costs. Nonetheless, one should keep in mind that for compliance to structural requirements the geometry of the structure could be changed.

For example, to increase the stiffness of a structure, one can select either a material with a higher stiffness, and/or create a stiffer geometry (shape/design). But, changing the stiffness of the material, represented by its Young's modulus, is not possible.

Likewise the density is a material constant. Other properties like the strength and the durability can be changed by (slightly) changing its composition (another alloy) or condition (temper).

In addition to these requirements, more requirements can be mentioned here that relate to the relation illustrated in Figure 3.2.

The manufacturability or workshop properties of materials relate materials to manufacturing aspects. To manufacture an aircraft, it is very important to have materials that have good workshop properties. For instance, aluminium alloys have good manufacturability, but titanium alloys don't. That means that processes like forming and machining (drilling, milling) are easy for aluminium alloys, but difficult for titanium alloys. In composites, glass and carbon fibre reinforced composites have good/adequate workshop properties, but aramid (Kevlar) fibre reinforced composites are very difficult to cut by machining operations, due to the very tough aramid fibres.

To emphasise the importance of the manufacturing aspects in relation to materials requirements, one should also consider that several manufacturing processes are relatively easy for one material, but impossible for other. For example, for manufacturing of a spar or stringer, extrusion and machining processes are available for metals, which are all inapplicable for fibre reinforced composites. Selection of the appropriate materials then relates to the available production processes.

Physical properties like electrical conductivity and the coefficient of thermal expansion (CTE) are important for specific features of the operation performance. The electrical conductivity of aluminium alloys make it easy to create a (safe) Faraday cage of the aircraft fuselage. For composites this is more difficult; sometimes extra strips or conductive meshes are required for this protection against lightning strike. In this respect, the CTE is also important because aircraft operate between +80°C (a hot day on the airport) and -60°C (at cruise altitude). Large differences in values of CTE of applied materials could cause extra problems, like the thermal stresses explained in the previous chapter.

Therefore, meeting the requirements should be achieved both on a structural level and material level. Once dominant material requirements are met, discrepancies could be solved on a structural level. For example, the earlier mentioned differences in CTE could induce thermal stresses in a structure. This cannot always be solved by changing one of the applied materials. The structural design solution, i.e. type of joint, direct contact between materials or separation by intermediate layers, could solve those specific issues.

CHAPTER-4
MANUFACTURING

4.1 INTRODUCTION

The previous chapter provided a brief description of the different material categories relevant for the aeronautical industry, together with some characteristics and typical applications. As mentioned there, the performance of a structure is to be evaluated by means of trade-off between material aspects (discussed in previous chapter), structural aspects (discussed in the next chapters), and manufacturing aspects.

This means that despite being dealt with in individual chapters in this book, these aspects relate directly to each other. Considering material aspects together with structural aspects, but without addressing the manufacturing aspects will never lead to optimized structures.

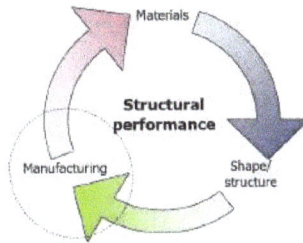

Figure 4.1: Illustration of the topic of interest in this chapter (Alderliesten, 2011. 4-1. jpg. Own Work.)

This chapter provides an overview of manufacturing techniques adopted in industry for manufacturing structures, components and products made of the materials presented in the previous chapter. This chapter limits itself primarily to the manufacturing aspects of metallic and composite materials, because they are considered most relevant for the aerospace industry.

4.2 METALS

In general, three different manufacturing categories can be identified for metallic materials to create a component or product

- ⦿ Casting
- ⦿ Machining
- ⦿ Forming

4.2.1 Casting

Casting is a production process in which liquid is poured into a mould containing the shape of the product. The shape of the cavity inside the mould defines the outer shape of the product. Although the principle is very old (it has been applied for over 5000 years), it is still often used for manufacturing complex parts. There are more materials that can be casted, like for example polymers, concrete and clays. The metals are heated above the melting temperature to provide a liquid that is subsequently poured into the mould. After cooling the part, the part is often retrieved by breaking the mould. An illustration of metal casting and of a typical product is given in Figure 4.2.

Figure 4.2: Casting: process (left) and typical product (right). Derivative from left: Marpockstudios, (2017), CC0 and right: CM_Photo (2016), CC0.

4.2.2 Machining

Machining is a solid state cutting and milling processes. This means that the process is applied to materials in their solid state at room temperature, which means that heating or melting of the material is not necessary. However, machining often required the use of a cooling agent, because cutting and milling create in itself heating of the material. The principle is illustrated in Figure 4.3. This figure illustrates the milling and cutting process. The first is a process that removes chips of material (milling, drilling, grinding, etc), while the latter is a does not remove chips, but separates (shearing) materials.

Machining is commonly used to produce large components for which small geometrical tolerances are required. Geometrical tolerances are defined as the maximum variation that is allowed in form or positioning of the product or component. Because the material can be fixed into its position in its solid state, machining can be performed with great accuracy, thus enabling small geometrical tolerances.

There are several types of tolerances possible:

- ◉ Form control
- ◉ Flatness
- ◉ Positioning
- ◉ Perpendicular or parallel

Although the machining process usually requires large and expensive equipment, the process is considered to be inexpensive for large quantities or large components, because it is relatively easy to automate. Especially high-speed machining enables production of components at high speeds with high accuracy and small tolerances in a relative short amount of time. Even thin-walled components can be easilymanufactured with high accuracy.

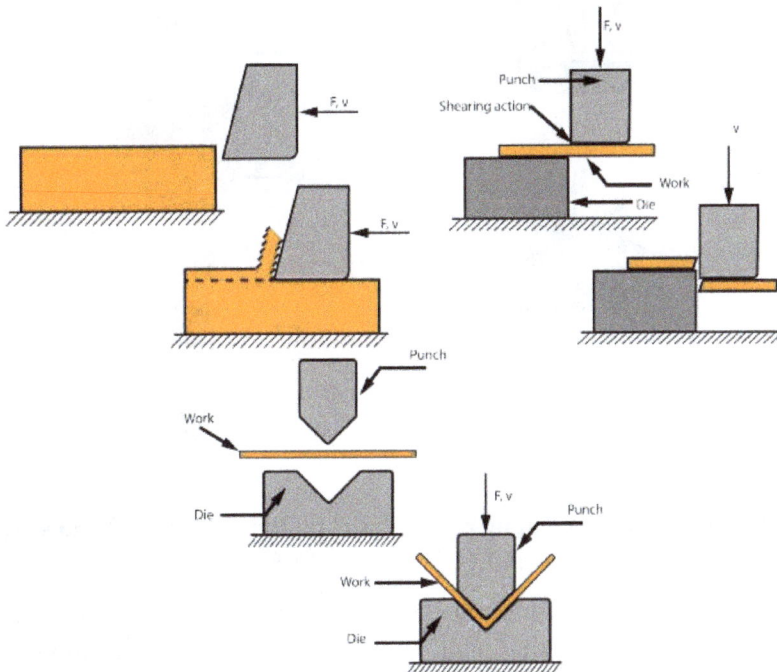

Figure 4.3: Illustration of the principles of machining (upper) and forming (lower). (TU Delft, 2018)

4.2.3 Forming

The category of forming, of which an example (sheet bending) of the principle is provided in Figure 4.3, can be further divided into forming of bulk material and forming of sheet material. Examples of the first sub-category are

⊙ Extrusion

⊙ Forging

Examples for the second sub-category are

⊙ Bending

⊙ Deep drawing

⊙ Roll bending

The forming principles of these processes are illustrated in Figure 4.4.

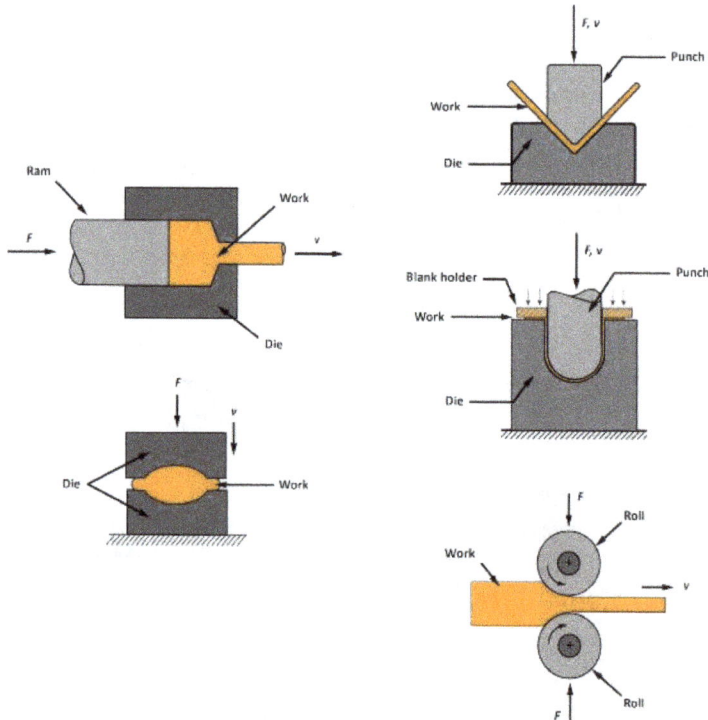

Figure 4.4 Illustration of the forming principles of bulk material (left) and sheet material (right). (TU Delft, 2018)

An important aspect to forming is that in order to change the shape of the material permanently, the material is plastically deformed. However, before plastic deformation occurs, the material deforms elastic until it reaches the yield strength of the material. The elastic deformation is reversible, which means that after the maximum deformation is reached and the forces are removed a small portion of the imposed deformation will be eliminated.

Consider sheet bending as illustrated in Figure 4.5. At the maximum bending deformation, the material has first gone through elastic deformation and subsequently plastic deformation. Thus once the bending force is removed, the material will spring back corresponding to the elastic deformation.

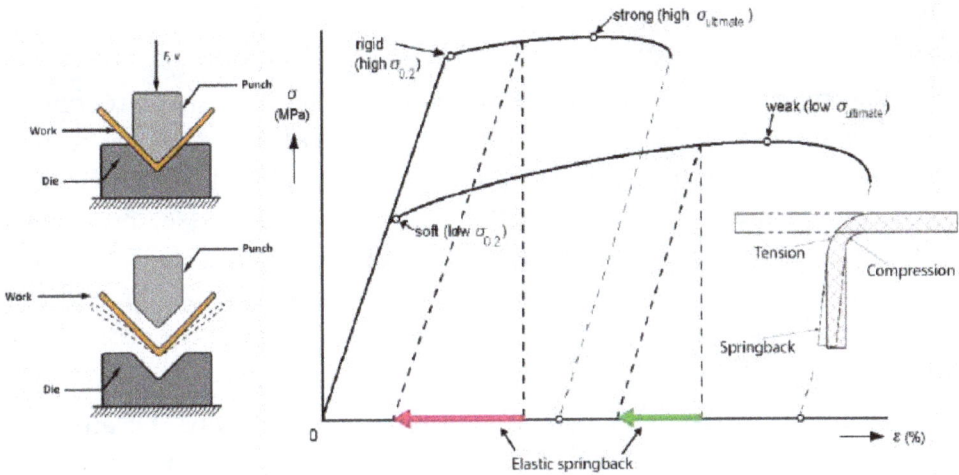

Figure 4.5: Illustration of spring back in sheet forming and relation to strength. TU Delft (2018)

There are two important aspects to this spring back. First, the amount of spring backis dependent on the level of the imposed stress. As illustrated in Figure 4.5, the spring back will be higher for higher yield strength, because higher yield strength implies more elastic deformation.

Here, one may observe the relation between material and manufacturing aspect in relation to the structural performance; often a high yield strength is preferred for a structure, to avoid permanent deformations at maximum operational loads. However, from a manufacturing perspective, high yield strength implies large spring back that needs to be accounted for.

That relates to the second aspect of spring back: tolerances. The final shape of the product is obtained after the forming forces are removed and elastic spring back has occurred. If small tolerances are required, this implies that the spring back must be known as accurate as possible. For the example of bending over a single line as illustrated in Figure 4.5, this may be very well achievable, but if the deformation is applied to double-curved shapes (3-dimensional), this becomes a complex calculation.

4.2.4 Forging

Forging is a forming process applied to bulk material. Special attention is given here to the production process. The interdependency between the material, manufacturing and geometrical aspects, illustrated in Figure 4.1, is clearly visible in the forging process.

Forging is a process where bulk material is deformed into another shape by applyingcompressive forces to the material. This process is be applied at elevated temperature. Obviously, at higher temperatures materials are easier to deform, because their resistance to deformation is lower.

The mould used to form the material into its new shape, defines with its inner contour the shape of the product. However, small tolerances cannot be achieved with this process, because the work applied to the material induces internal stresses. Together with little elastic deformation (see previous section), these residual stresses will find equilibrium while settling to its final shape after unloading. Because the level of detail achievable with forging is limited, machining is commonly applied after the forging process, to achieve the required geometry with the desired accuracy and tolerances.

However, this requires some additional considerations, because if internal residual stresses have obtained equilibrium after the forging process, this may imply that once material is removed with machining the equilibrium has changed and subsequently the shape. Therefore, either thorough analysis of the formation of residual stresses is required to predict potential 'spring back' after machining, or machining has to be applied carefully by removing in a symmetric manner and in small amounts. An example of a forged rid is illustrated in Figure 4.6.

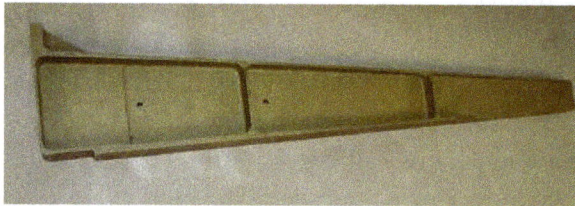

Figure 4.6: A forged rib of a control surface (Saunders-Smits, 2018)

The example visualized in Figure 4.6 illustrates another relation shown in Figure 4.1, because the selection of the production steps in manufacturing a component, may imply that the residual stresses, that are in equilibrium in the product in its final shape, may or may not be significant. The lower the residual stresses in the final product, the less issues they may induce on the structural performance.

4.3 COMPOSITES

In manufacturing of composite structures, several production techniques can be identified that can be classified in three groups:

- Placement of fibres in dry condition
- Placement of fibres in wet condition
- Placement of fibres after pre-impregnation

4.3.1 Filament winding

Filament winding is a production technique that can be applied by placing the fibres either in dry or wet condition on a mould with a given geometry. Because of the rotational movement during winding, typical products are products containing a cylindrical geometry, like for example pressure vessels. Open geometries such as, for example, bath tubs cannot be made using filament winding. Products made using filament winding often have a textile appearance.

Characteristic for the filament winding process is that the placement of fibres is bound by the initial orientation and friction, with no free variable orientations possible.

The mould used to define the shape of the product over which the fibres are wound is called a mandrel. There are two options that can be applied for the mandrel:

⊙ Removable mandrel; the mandrel should be solvable, collapsible or tapered in order to be removed

⊙ Used as liner; the mandrel becomes part of the final product and serves for example as liquid and gas tight inner liner.

The windings that are possible are given by:

⊙ Hoop $\alpha_i = 90°$

⊙ Helical $\alpha_i = \pm \eta°$

⊙ Polar $\alpha_i \approx 0°$

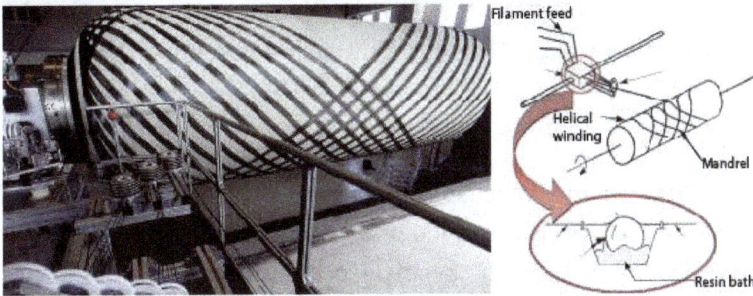

Figure 4.7: Filament winding machine (left) and a Schematic presentation of filament winding (right). Derivative from left: Gdipasquale1, (2018), CC-BY-SA4.0 and right: Esi. us1, (2001), Public Domain.

4.3.2 Pultrusion

Pultrusion is a process equivalent to extrusion, see Figure 4.4, but where the material is pulled through the mandrel rather than pushed. The material is pulled trough a mandrel that defines the shape of the product. Although the reinforcement with fibres is possible and commonly applied, there are limitations to the way fibres can be positioned in the product. This limitation is related to the process. The

reinforcements that are possible are

- ◉ Rovings, strands or unidirectional material
- ◉ Chopped strands (fibre mats)
- ◉ Woven fibres

The first category is used the most. In any case, whatever category is selected, a certain amount of continuous fibres is necessary to provide the strength in longitudinal direction for pulling. For the matrix material thermoset polymers are applied, such as polyesters and vinyl esters.

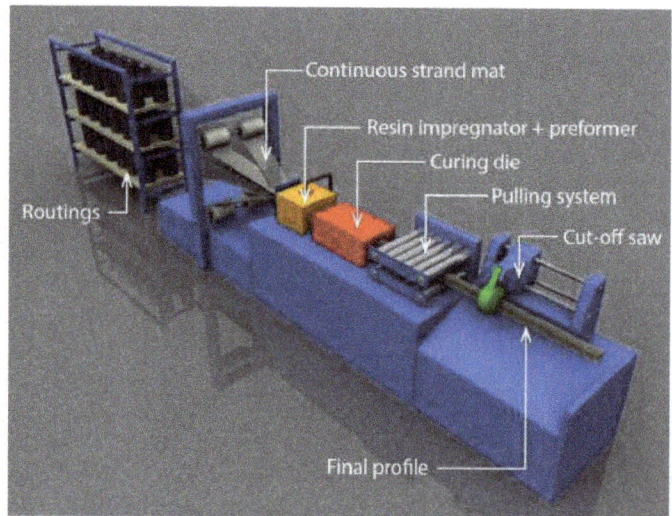

Figure 4.8: Schematic presentation of a pultrusion machine. Derivative from Arnd, (2005), CC-BY-SA3.0.

4.3.3 Lay-up

Lay-up is a manufacturing process that can either be applied manually, or fully automated. Manual lay-up is often used for small quantities of a product. Because of the relative low quantities of aerospace components (except for single aisle aircraft, most commercial aircraft are manufactured at a rate that is closer to prototyping than full scale industrial production) manual lay-up is often applied. However, with the introduction of composite fuselage structures (Boeing 787 and Airbus A350) the need for automation of the production process increases. This is especially the case, because circular shaped barrel components cannot easily be made with manual lay-up. The automated lay-up process is typically referred to as tape laying or fibre placement.

Comparing tape laying with manual lay-up, some benefits and drawbacks can be identified. Because automation implies numerical control, the placement usually is more accurate and more repeatable than manual lay-up. As a consequence the difference between identical parts is less for tape laying. For fibre reinforced composites, the improved accuracy in placement of fibres results often in better mechanical properties.

However, the apparent drawback is that it requires an expensive tape laying machine, which makes the automated process even more expensive. As a consequence, only large volumes or high end products, like for example space technology, justify the high investments.

In general, components made by lay-up can be very large but are limited by the time needed for impregnation and curing and the size of the autoclave. An autoclave is a pressure oven in which components can be cured at elevated temperature and pressure. For example, the fuselage barrels of the Boeing 787 require, because of their diameter, a very large autoclave.

Lay-up can be applied both using dry lay-up, which is then followed by impregnation and subsequent curing, and wet lay-up, which is only followed by curing.

A typical variant of lay-up is prepregging. This process uses pre-impregnated material (i.e. fibres that are impregnated with resin and subsequently consolidated as tape or prepreg), which is positioned on a mould either by manual or automated lay-up. After lay-up, the component is placed underneath a bag that is put in vacuum to remove the air from the component, before it is cured in, for instance, an autoclave. This lay-up variant enables manufacturing components with high quality, but in return implies a rather expensive process.

A quicker and cheaper process uses a spray up technique. Instead of long continuous fibres short chopped fibres (10-40 mm) are applied that are sprayed onto the mould. As a consequence, the fibres are positioned in a random orientation, which provides in-plane isotropic properties.

The disadvantage of this process is that it results in components with low mechanical properties compared to lay-up using continuous fibres, and the quality depends on craftsmanship. This implies that repeatability of this process is low from one worker to another.

4.3.4 Resin transfer moulding (RTM)

Resin transfer moulding is a manufacturing process where dry fibres are placed in a stiff and rigid mould, which is subsequently closed. With the use of pressure difference (i.e. high pressure at the entrance of the mould and low pressure at its exit) resin is drawn through the mould cavity impregnating the dry fibres. After this process step is completed, the impregnated component can be cured. The principle of the RTM process is illustrated in Figure 4.9.

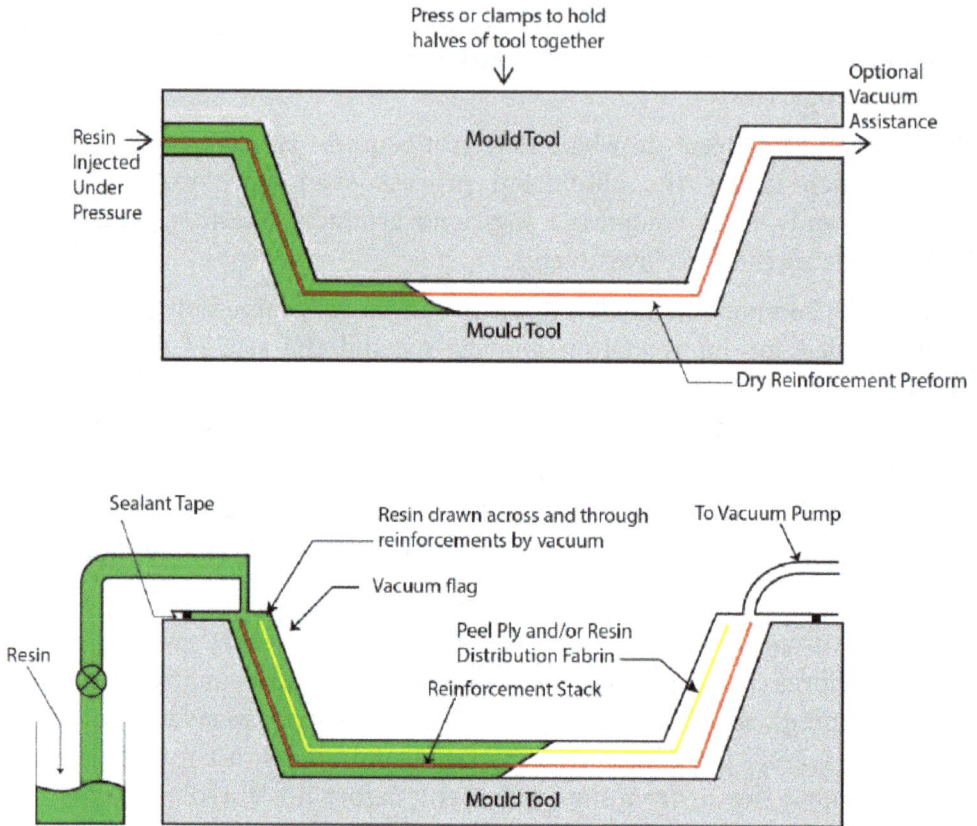

Figure 4.9: Principle of resin transfer moulding and vacuum infusion. (TU Delft, 201)

An alternative process, but closely related to RTM is vacuum infusion. In that process, dry fibres are also placed in a stiff and rigid mould, but the mould is closed with a flexible film. Putting the dry fibres underneath the film under vacuum will compress the material and pull the resin through the dry fibres in the mould.

The third variant is the vacuum assisted resin transfer moulding (VARTM). This process is in general similar to RTM, but the pressure differential is not only created by applying pressures higher than 1 bar at the injection side, but also by applying vacuum at the exit side.

The difference between the pressures used in the three processes is summarized in Table 4.1. Typical examples of applications manufactured with vacuum infusion are given in Figure 4.10.

Table 4.1 Typical pressures applied to obtained requires pressure differential

Pressure	P_1 (outside)	P_2 (inside)	ΔP
Resin Transfer Moulding	>1bar	1bar	>1bar
Vacuum Assisted Resin Transfer Moulding	>1bar	<1bar	>1bar
Vacuum infusion	1bar	<1bar	~1bar

Figure 4.10: Example applications of vacuum infusion: Contest 67CS (upper row) and Eaglet rudder (lower row). Derivative from top left: Copyright 2016 by Contest Yachts, Reprinted with Permission.; Top middle and top right: Copyright 2006 by Lightweight Structures B.V., Reprinted with Permission; bottom row: TU Delft, (N.D.).

4.4 THERMOSET VERSUS THERMOPLASTIC

With respect to composites manufacturing, some additional remarks shall be made. Traditionally, composite applications for aerospace structures were manufactured using thermoset resins, like for example epoxy and polyesters. However, the introduction of thermoplastic polymers as matrix material for composite structures has implications on the production aspects.

A qualitative comparison between the characteristics of thermoset and thermoplastic components and their manufacturing is given in Table 4.2.

In general, the advantage of thermoplastic composites is that they enable rapid manufacturing, producing components that can be recycled (melting the material again), welded (locally melting the material and cooling after components are pressed against each other), and that the process does not depend on chemical reactions. Compared to thermoset resins an important disadvantage is the required high temperatures and pressures, the limited storage life and out-gassing characteristics.

Aspects that must be considered before selecting the application of thermoplastic resins is that the melt processing used to produce thermoplastic composites limits the viscosity that can be achieved. A dimensional limitation due to the viscosity requirements can be illustrated using the example of wind turbine blades. Because these blades are very large and long, it is impossible to push the resin through the full length of the components even when applying high pressures. Therefore, the melt process of thermoplastic composites is considered an unsuitable process for manufacturing wind turbine blades.

Table 4.2 Comparison between thermoset and thermoplastic

Aspect	Thermoset	Thermoplastic
Material	Liquid components A and B	Single solid matrix
Melting step	No	Yes
Impregnating fibres	Yes	Yes
Chemical reaction	Yes	No
Material after cooling	Solid matrix	Solid matrix

CHAPTER-5

AIRCRAFT & SPACECRAFT STRUCTURES

5.1 INTRODUCTION

To be able to design and analyse aircraft or spacecraft structures, one first has to become familiar with the structural elements and their functions. In this chapter the different structural elements that can be observed in aeronautical structures and their functions are discussed. This explanation is primarily qualitative of nature, but will be further explored, where loads acting upon these structural elements are discussed and the stresses that they induce.

5.2 AIRFRAME

The aircraft consists of numerous different elements and components that each fulfill their own function. However, not all elements of an aircraft are considered to be structural elements or fulfill a load bearing function. To distinguish elements that areconsidered to be part of the structure of an aircraft or spacecraft from other elements, these elements are often referred to as *airframe*.

There are several definitions that can be given for the airframe. An airframe is

- The aircraft or spacecraft without installed equipment and furnishing
- The skin and framework (skeleton) that provide aerodynamic shapes
- The load bearing parts that take up forces during normal flight, manoeuvres, take-off, landing etcetera.
- The parts that together protect the contents from the environment

The reason that these different definitions exist can be attributed to the different types of airframes or structures that can be identified. For example the Fokker Spin, illustrated in Figure 5.1, is completely different in appearance than the military aircraft that are operated these days, see for example the illustration in Figure 5.2 of a Sukhoi SU-25.

Defining airframe as the parts that protect the contents from the environment may be considered applicable for the fuselage and wing structure of a commercial aircraft, but is hardly applicable to the Fokker Spin, where the pilot is only protected from the environment by elements that are not part of the airframe.

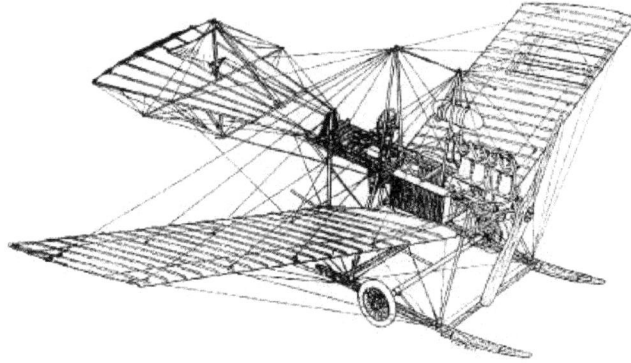

Figure 5.1: Illustration of the Fokker Spin or Fokker Spider. (Anon., N.D.)

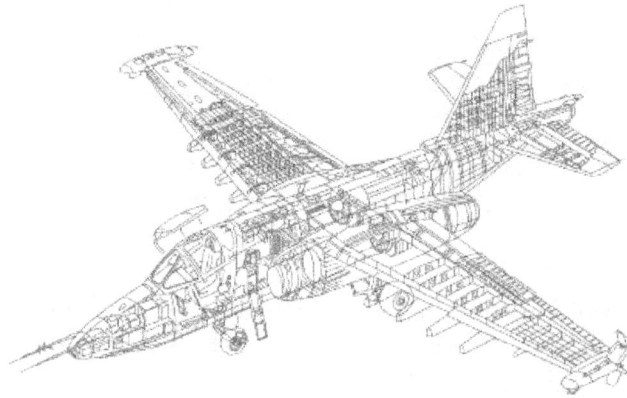

Figure 5.2: Illustration of a Sukhoi SU-25T Fighter Jet. (Altoing, 2011, CC-BY-SA3.0)

Within the airframe a more detailed distinction is made between structural elements that fulfil a critical or a non-critical function. Elements that fulfil a critical function are elements that in case of damage or failure could lead to failure of the entire aircraft or spacecraft. These structural elements are referred to as *primary structures*. Structural elements that fulfil non-critical functions are elements that carry only aerodynamic and inertial loads generated on or in that type of structure. These structural elements are called *secondary structures*.

Figure 5.3: Photo of the shell structure of the cockpit section of the Fokker 100 aircraft (component located in the collection of the Delft Aerospace Structures & Materials Laboratory – TU Delft, N.D.)

5.3 STRUCTURAL CONCEPTS

In the past century, different structural concepts have been developed and explored that form the basis of current structural design in both aviation and space. In this section these concepts will be described briefly.

5.3.1 Truss structures

The first structural concept applied in aircraft structures was the truss structure. This structural concept was widely applied at the time in, for example, (railway) bridge structures. Truss structures are made of bars, tubes and wires (wire-bracing of a structure). A famous example of wire-bracing of a structure is the Fokker Spin shown in Figure 5.1.

The aircraft in the first decades of aviation were built with truss structures in which the bars, tubes and wires carried all loads. The skin or fabric coverage did not contributeto the load bearing function.

The elements of the truss structure transfer the loads primarily via tension and compression. Structural rigidity is obtained with the diagonal function, i.e. wires or rods placed under 45°. In Figure 5.4 the diagonal function in the truss structure is illustrated. The concept can be applied in three-dimensions to sustain bending and torsional loading.

The diagonal function can be performed by either wires (only in tension), rods and tubes (tension and compression) and sheet material. The latter solution has led to another structural concept that will be discussed next.

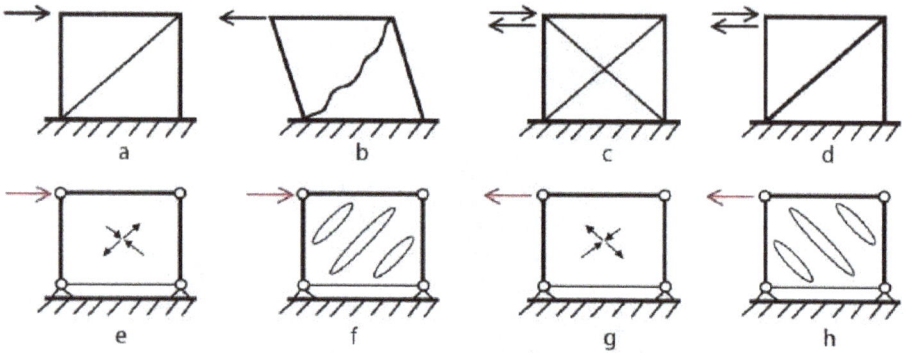

Figure 5.4: Illustration of truss structure concept: diagonal function with wire-bracing (a-c), with rods (d), and with sheet (e-h). (TU Delft, N.D.)

Although truss structures were the first structural concept introduced in airframes, this does not automatically mean that the concept has become outdated. In some modern applications, the introduction of tailored truss structures has led to significant weight savings.

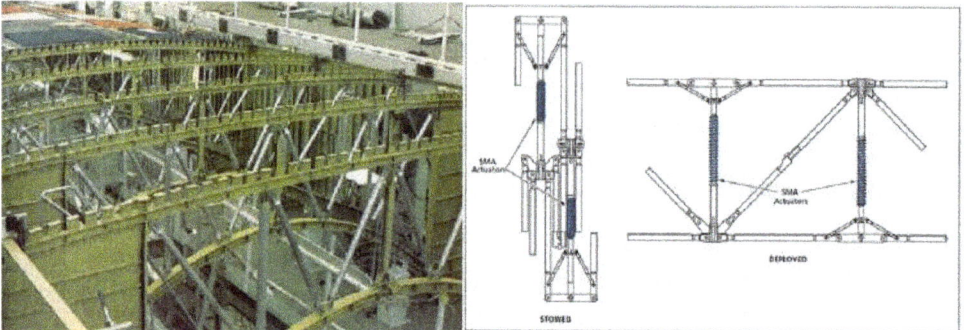

Figure 5.5: Examples of truss structures, left: Ribs in the centre fuselage section of the Space Shuttle using boron-aluminium tubes, (Rawal 2001), and right: deployable truss concept for space applications. Derivative from NASA (2001, 2010), Public Domain.

Such an example is illustrated in Figure 5.5, where the ribs in the centre fuselage section of the Space Shuttle have been constructed partly from a truss structure using boron-aluminium tubes leading to a 45% weight saving over the original baseline design. Truss structures also enable deployment concepts which are of interest for space applications.

5.3.2 Shell structures

With the introduction of metal in aircraft structures a different structural concept was introduced. This concept relates to the application of sheet material in a truss structure to fulfil the diagonal function. In this truss configuration, the sheet performs the diagonal function in shear, whereas the wires and rods only perform that function in tension and compression. For this reason, sheet material is very efficient to reinforce such structure. However, with the introduction of sheet material in the structure, it is no longer considered a truss structure, but rather a thin-walled *shell structure.*

Four functions that can be performed by sheet material are illustrated in Figure 5.6. In addition to tension and compression (also with tubes and wires), sheet material can carry load in shear (diagonal function), but can also seal a structure from its environment. This is particularly of interest for the pressurization of a structure (including the ability to carry membrane stresses) and for the application of integral fuel tanks.

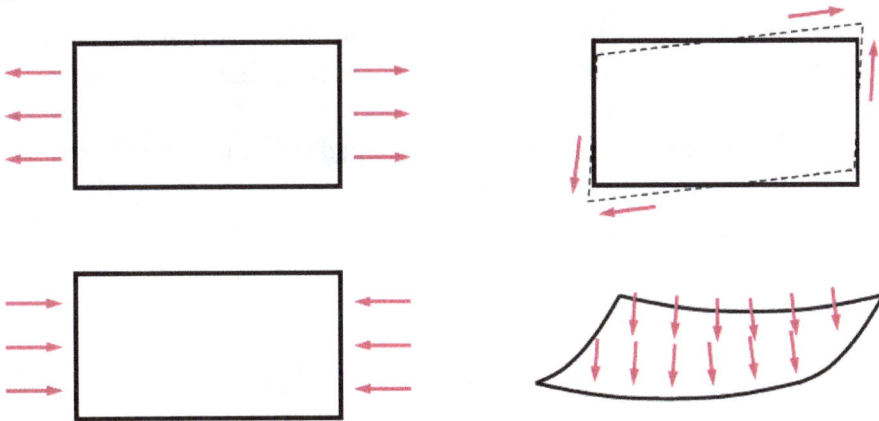

Figure 5.6: Four functions of sheet: tension, compression, diagonal/shear, sealing (air/ fuel tight). (Alderliesten, 2011)

To fully exploit the benefits of sheet material, it may be necessary to reinforce the sheet with other elements. Without any additional reinforcement, sheet material can easily carry tensile loads up to its ultimate strength. However, a sheet in compression will bulge out without being able to carry significant compressive loads, see Figure 5.7 (a). Only when the sheet is reinforced or geometrically stiffened, as illustrated in Figure 5.7 (b,c), it is able to carry more compressive load.

Apparently, the sheet stiffness cannot be used when it bulges, but when the sheet is formed into a profile its obtained stiffness can be used in compression. This means that stiffness in the structural context is not only a material property, but also a geometrical property. Figure 5.7 (c) illustrates that the sheet of example (a) can be stiffened by another sheet that is formed into a profile, where the combination of the two sheets becomes a load carrying structure in which the flat sheet carries a significant part of the applied load.

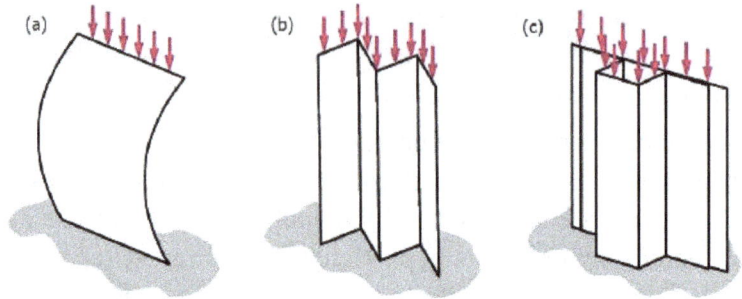

Figure 5.7: Stiffness of sheet material versus geometrical stiffness. (Alderliesten, 2011)

The combination of a flat or slightly curved sheet with stiffeners is called a shell structure. There are multiple stiffener geometries that can be used to reinforce a shell structure. The difference between a sheet reinforced with so-called hat stiffeners and a sheet reinforced with an L-shaped stiffener is illustrated in Figure 5.8. The stability of the hat-stiffener is higher than the L-stiffener, where at certain compressive loads the flange will bulge out, similar to the flat sheet in Figure 5.7 (a), but only at the unsupported edge.

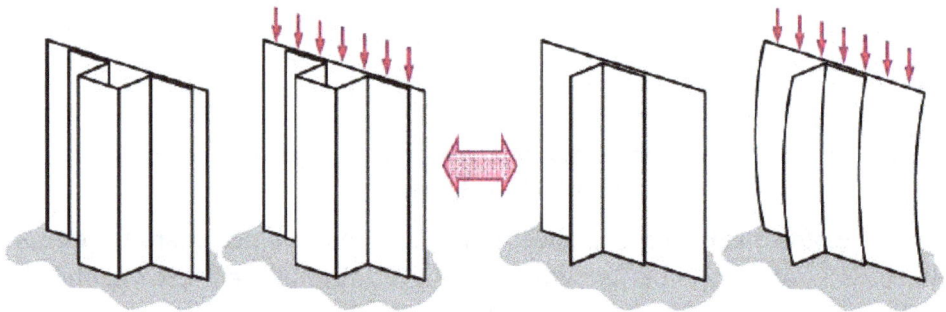

Figure 5.8: Effect of stiffener geometry on deformation induced by panel compression. (Alderliesten, 2011)

For this reason the shape of the stiffener is considered in structural design. Heavy loaded structures, such as, for example, the upper wing panels (upward wing bending causes compression in the upper shell structure) hat-stiffeners are often applied. In shell structures that carry less compressive loads (lower wing panels), often L- shape or Z-shape stiffeners are applied. An additional advantage is that these stringer shapes enable inspection from all sides, while hat stiffeners cannot be inspected internally.

Another design consideration is the distance between the stiffeners. A single stiffener attached to a flat shell (Figure 5.9) has only a limited and local effect on the stability of the shell structure. Depending on the stiffener geometry an effective width can be determined. Based upon this effective width, spacing between stiffeners must be chosen in the design.

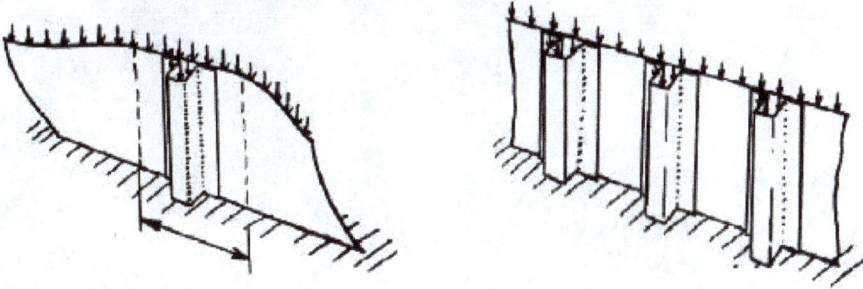

Figure 5.9: Effect of stiffener spacing on stability of panel under compression. (TU Delft, N.D.)

Where initially a beam structure was designed as truss structure, where bending was taken up by normal forces in the upper and lower tubes and the diagonal function performed by the diagonally placed elements, the introduction of sheet material has led to more efficient beam concepts. The sheet performing the diagonal function is called the web plate, whereas the elements taking up the (bending) loads by normal forces are called the girders. This beam concept is illustrated in Figure 5.10.

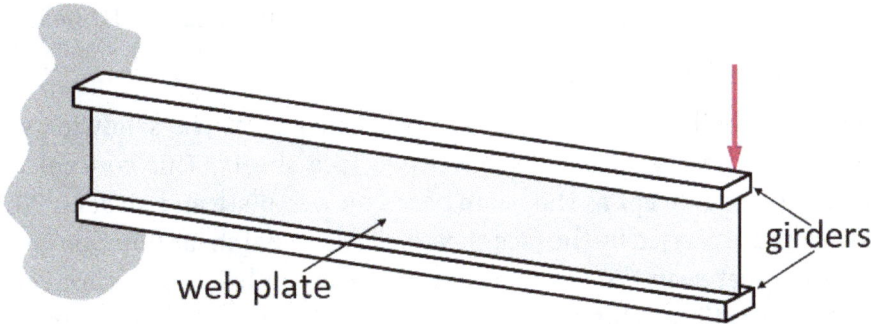

girders

web plate

Figure 5.10: Concept of a beam based on sheet web with girders. (Alderliesten, 2011)

An illustration of the application of the beam concept in a fuselage structure is given in Figure 5.11. Here the frames, clips and stringers all utilize the characteristic beam concept with web plate and girders.

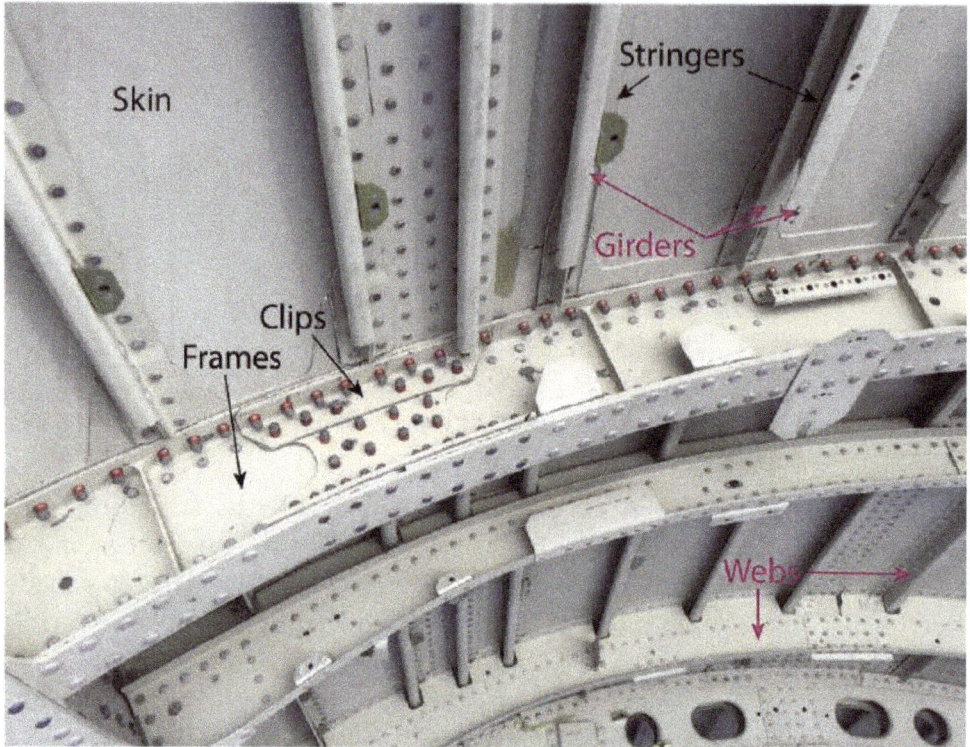

Figure 5.11: Frames and stringers based upon the beam concept. (TU Delft, N.D.)

5.3.3 Sandwich structures

An alternative shell structure is the sandwich structure. The sandwich structure contains a low density core in-between two face sheets. This concept basically exploits the same concept as the beam based on web plate and girders. Where the normal forces are carried by the face sheets, the diagonal shear function is provided by the core in-between the facings. However, the sandwich structure carries the loads in two directions. This similarity in concept is illustrated in Figure 5.12.

Sandwich structures are in fact bonded structures because of the adhesion between face sheets and core. There are several core concepts that could be applied. The honeycomb core structure illustrated in Figure 5.13, could be made from either metals or polymers. In composite sandwich structures, polymer honeycombs are being used, while in the past (see Bréguet Atlantic example on the next page) aluminium sandwiches structures used aluminium honeycomb cores. The advantage of honeycomb cores is the low weight due to the high percentage of empty volume.

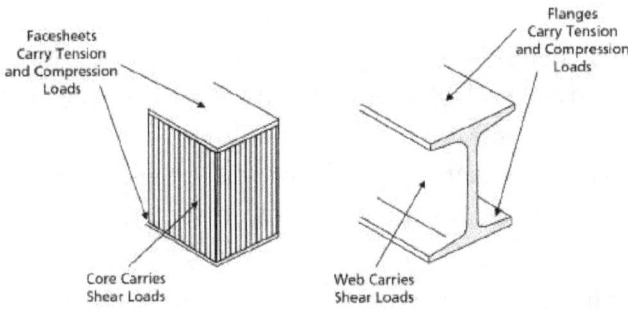

Figure 5.12: Illustration of similarity of concept between beam and sandwich; the girders and face sheets carry the normal forces, while the web plate and core provide the diagonal function. (TU Delft, N.D.)

Alternative core solutions are based on the application of balsawood or foams. The foam cores can either be made of polymers or metals.

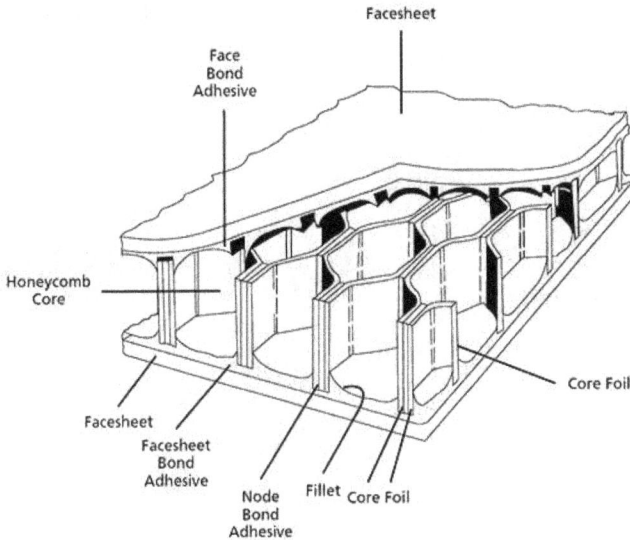

Figure 5.13: Illustration of sandwich concept utilizing the honeycomb core. (TU Delft, N.D.)

The advantage of the sandwich concept is the inherent high bending stiffness. For this reason, sandwich structures are often applied as floor concept in passenger and cargo areas of aircraft. To illustrate the benefit in bending stiffness and strength, in Figure 5.15 for a given sandwich lay-up the relative increase in bending stiffness, strength and weight is shown. With adding low density core material, the weight is only slightly increased, while the bending stiffness may increase order of magnitudes.

Example: Sandwich panel concept

The sandwich structure may be considered an excellent structural concept, because it requires no additional stiffening elements providing a smooth surface.

However, examples from the (recent) past have also put the emphasis on the drawback of this concept.In 1970 the Bréguet Atlantic, shown in Figure 5.14 (a) revealed substantial corrosion issues related to moisture absorbed by the sandwich structure. One may assume that the sandwich structure is a closed system, but nonetheless, moisture is being picked up by the sandwich structure, and the moisture remains subsequentially trapped inside, see the X-ray photo in Figure 5.14 (b).

To illustrate that this is not a particular problem related to metallic sandwich structures, but to sandwich structures in general, another example is shown in Figure 5.14 (c). On March 6, 2005, an Airbus A310-308 operated by Air Transat designated Flight 961 had to perform an emergency landing after it lost in-flight a composite sandwich rudder of several meters length.Subsequent investigation revealed that moisture absorbed by the sandwich panel trapped between core and face sheets caused rapid delamination of the face sheets from the core, as a result of volumetric increase of the frozen moisture.

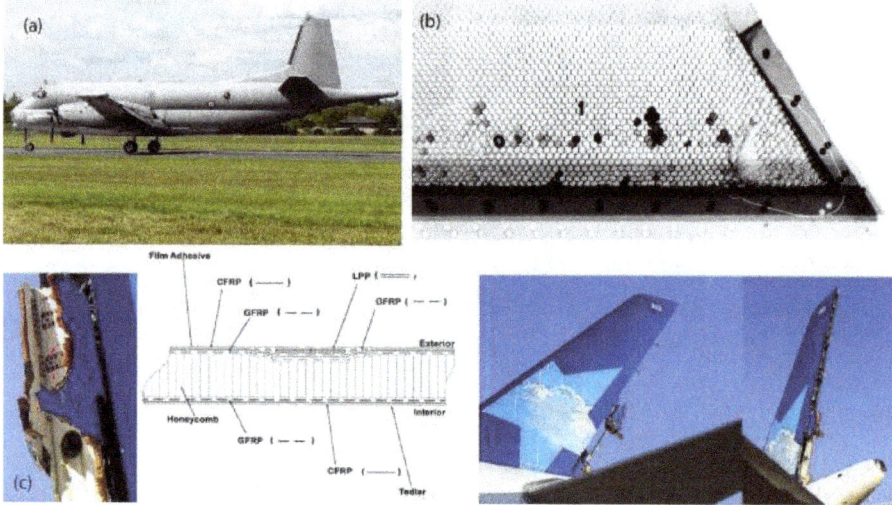

Figure 5.14: Illustration of detrimental environmental influence on sandwich concept: corrosion in metallic sandwich (a,b) and rapid delamination in composite sandwich (c). Derivative from top left: Bidini, (2012), CC-BY-SA3.0; top right: TU Delft, (n.d.), 5-14-b.jpg, Own Work; bottom row: TSB, (2007), Copyright TSB, Use Permitted.

In general, the sandwich concept is an excellent structural concept. However, one should bear in mind that there are several drawbacks related to this concept. First of all, details like the mechanical joints to other structural elements are usually complex and expensive. Simple riveting or use of bolts is not possible, because the soft core is not able to carry the bearing load induced by bolts, causing delamination between facing and core. The high clamping forces often required for bolted joints is also not possible, because the out of plane stiffness of the sandwich structure is only related to the local stiffness of the face sheet and therefore low.

The second drawback is related to durability and aging of sandwich structures. As explained in the example earlier, the environmental aspect of moisture absorption by the sandwich panels is significant. Experience with for example composite sandwichlanding gear doors on Airbus aircraft has revealed that over a few years the weight of such panel may increase by a kilogramme due to the high amount of moisture absorption. Aside from the effect of increased weight on the aircraft performance; it may also impair the structural integrity of the aircraft, similar to the example of the Air Transat aircraft discussed before.

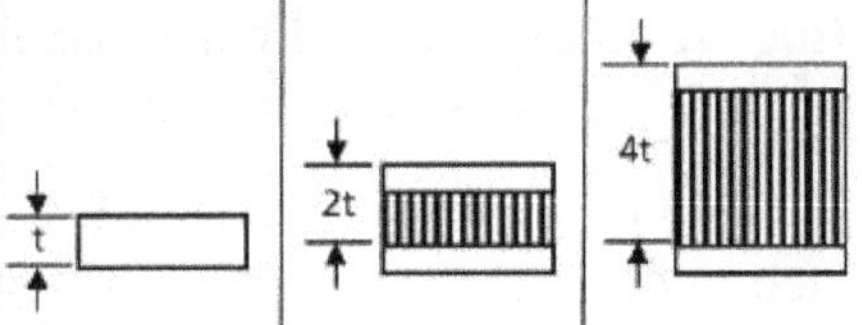

	Solid Material	Sandwich Construction	Thicker Sandwich
Stiffness	1.0	7.0	37.0
Flexural Strength	1.0	3.5	9.2
Weight	1.0	1.03	1.06

Figure 5.15: Effect of core thickness on bending stiffness. (Alderliesten, 2011.)

5.3.4 Integrally stiffened structures

The stiffened shell structures can be manufactured by attaching the stiffening elements to the face sheet; this can be done by either using mechanically fastening (riveting, bolting) or by bonding. Another method of manufacturing stiffened shell structures is by taking a thick plate and machining away the material between the stiffeners. This is illustrated in Figure 5.16. The advantage of the integral concept is the relative low cost of production. Machining, or nowadays high-speed machining, is performed automated and computer controlled, which enables high volume production at relatively low production costs. A second manufacturing benefit of integral structures is the low amount of parts. This reduces the logistic burden in a production facility.

Obviously, this concept leads to a high amount of scrap material. However, that does not add to the manufacturing costs, because nowadays, the metal suppliers are pre-milling such integral components before shipping them to the aircraft manufacturer. The highest amount of scrap metal remains at the metal supplier who can insert that material directly back into their production line.

Another advantage is that with machining the thickness can be tailored continuously, which leads to optimized structures with respect to weight. However, a drawback related to machining is that only simple blade stiffeners can be applied. Complex stiffener concepts cannot be manufactured due to the accessibility of the milling heads.

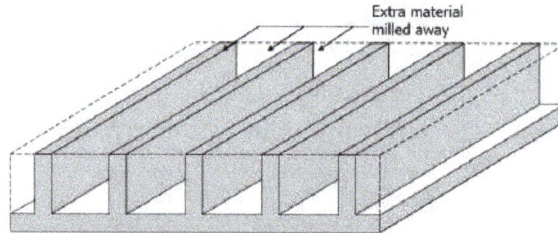

Figure 5.16: Creating an integrally stiffened panel by machining a thick plate. (Alderliesten, 2011)

The difference between the earlier methods of creating a stiffened panel and an integrally stiffened panel is that the latter consists of a single component. Or to put it the other way around, the earlier mentioned concepts of riveting or bonding such panels are build-up structures.

This means that in case of damage or cracks in the structure, the build-up structure contains natural barriers that stop or retard the crack until a new crack has initiated in the subsequent elements. In case of an integral structure, there is no natural barrier, which means that the crack will grow continuously through the panel. As a result, the integral structure is considered less damage tolerant (i.e. lower ability to withstand damage and damage growth) than the build-up structure.

Because this concept is most cost-effective for thick reinforced plates, like for example lower wing panels, the integral machined panel concept is mostly used on larger aircraft.

5.4 FUSELAGE STRUCTURES

The fuselage structure of pressurized aircraft is in fact a thin-walled pressure vessel, exploiting the stiffened shell concept. In a fuselage structure different elements can be identified that together provide strength and geometrical rigidity. These elements are:

- Fuselage skin
- Frames
- Stringers
- Bulkheads
- Splices and joints

The longitudinal stiffness of the structure is created in part by the closed cylindrical shell structure, but to great extent by the application of stringers to the skin panels. Another concept to create that rigidity is by combining stringers into longerons (i.e. heavy longitudinal stiffeners that carry large loads). This concept then has only few longitudinal stiffeners that have a larger cross section. An illustration of such concept is given in Figure 5.17.

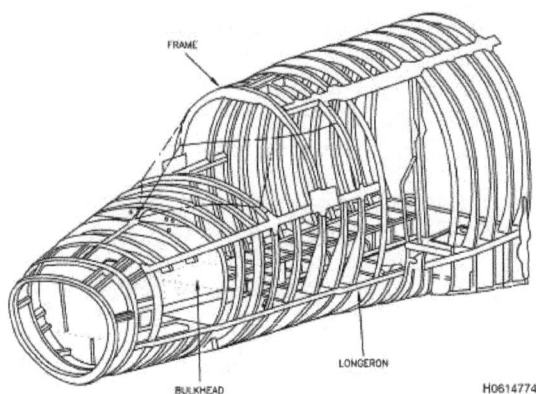

Figure 5.17: Longitudinal stiffness created with longerons.
(Anon., N.D.)

Pressurized fuselages require the application of bulkheads at the front and rear end of the fuselage to create an air tight pressure vessel and to maintain the forces related to the pressurization.

5.5 WING STRUCTURES

Wing structures contain similar stiffened shell concepts as explained in section 5.3, but due to the different functions to be fulfilled, the elements have a different appearance. In general, wing structures can be divided into:

◉ Wing skin panels

◉ Ribs

◉ Spars

Where the assembly of spars and skin require some further elaboration.

A first glance at wing structures reveals that, despite the use of similar elements, a certain design freedom results in different structural appearances. This is illustrated in Figure 5.18 where four structural wing lay-outs are compared.

Some wings contain straight spars from wing root (wing-fuselage connection) to wing tip, while other wings contain kinks in the spars. Such a kink in the spar most often has to be achieved by connecting two spar elements at that particular location.

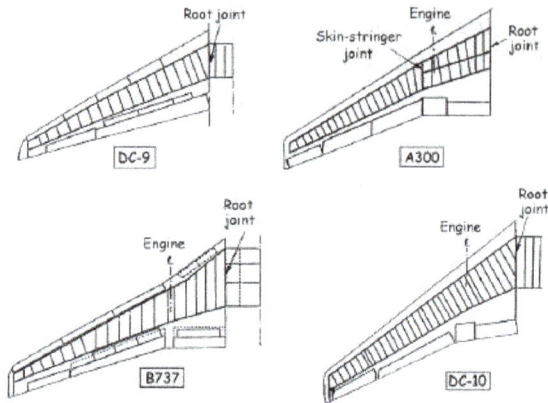

Figure 5.18: Comparison of four wing structural lay-outs. (TU Delft, n.d., 5-18.jpg. Own Work.)

Another aspect that strikes the eye is the orientation of the ribs in the wing structure.Where some wings contain ribs that are placed in flight direction, other wings contain ribs that are placed more or less perpendicular to the spars and wingspan direction.

In general, wing structures are characterised by the significant length the structural elements may have; the wing skin panels and the spars are generally made out of single components if possible. To get an indication of the length of large wings structures, in Figure 5.19 a photo of the A380 wing structure is shown. Because of the length aspect and the logistics related to that, some aircraft have wing structures that are sectioned into, for example, inboard and outboard wings, or centre wings and outer wings, like for example the C-130 Hercules illustrated in Figure 5.19.

Figure 5.19: Left wing of the Airbus A380 and wing structure of a C-130 Hercules (right). Derivative from left: Dr. Brains (2013), CC-BY-SA 3.0; and right: SAP – U.S. Air Force, (2007), Public Domain.

Despite the advantages of sectioning large structures with respect to manufacturingand assembly, a potential disadvantage may be related to the highly loaded joints asa result. For the C-130 Hercules, this joint, often referred to as the rainbow fitting, is a well-known point of concern.

5.5.1 Ribs

The ribs in wing structures provide several functions. First of all, they maintain the aerodynamic profile of wing. Because the aerodynamic profile of the wing is defined in flight direction, this may be an important reason to place the ribs in flight direction.

However, aside from the manufacturing challenges to place the ribs under such angles with the spars that are positioned in wing span direction; there is another obvious reason why placing ribs perpendicular to the spars is preferred.

The length of the ribs is shorter when placed perpendicular to the spars and wing span direction, than when placed in flight direction. The rib placement may be chosen such to minimise weight.

The second function of ribs is to transfer the aerodynamic and fuel loads acting on skin to the rest of the wing structure. The aerodynamic loads are related to the under- pressure above the upper skin, while the fuel loads act directly on the lower wing skins. These loads are brought into the structures via the ribs.

Thirdly, the ribs provide stability against panel crushing and buckling. This is illustrated in Figure 5.20; without any ribs, the upward bending of the wing would crush the upper and lower wing panels. With ribs at certain length, the upward bending would cause buckling to the upper wing skin, but keep the lower wing skin at distance of the upper wing skin. With the optimal rib spacing, both crushing and buckling are prevented and the shape of the wing is preserved.

A fourth function of the rib that can be identified is the introduction of local loads. Aside from aerodynamic and fuel loads, the wing structure is locally loaded by the landing gear during landing, taxiing and take-off, and, during flight, the engines, flaps and ailerons will locally apply loads to the structure as well. These loads are locally introduced into the structure by the ribs.

A fifth function is the sealing function in case integral fuel tanks are used in the wing structure. The ribs prevent the surge and splashing by sectioning the fuel tanks in individual bays.

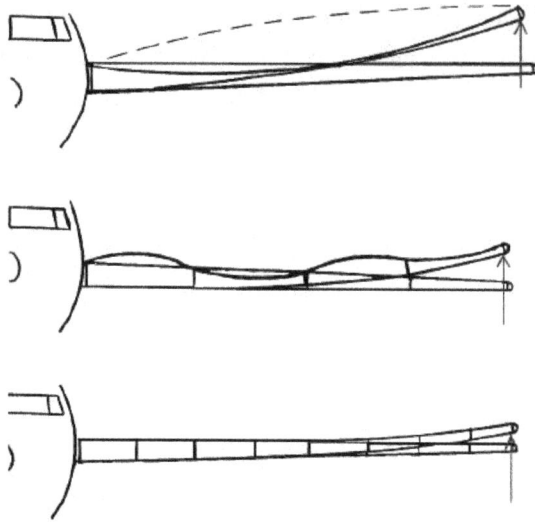

Figure 5.20: Illustration of the rib function; stability against crushing and buckling. (TU Delft, N.D.)

Especially the local loads introduced into the wing by the ribs imply that different loads are to be carried by individual ribs. As a consequence, different design concepts or design philosophies may be chosen for individual ribs in the wing structure. This is illustrated with the centre wing box of the C-130 Hercules in Figure 5.19; where, depending on the location in the wing box, different rib designs have been applied.

In general, the selection of rib type and related manufacturing method depends on;

- Loads
- Design philosophy
- Available equipment and experience
- Costs

The loads relate to the functions of a particular rib, i.e. does it only provide stability against buckling and crushing, or are significant loads being introduced locally by, for example, the landing gear? The design philosophy determines the magnitude of the stresses that are allowed in rib structures.

The available equipment and experience is an important aspect in this case. If certain equipment or machines are not available, it is most likely that the rib design that requires this equipment will not be chosen. To some extent it relates to the fourth aspect, costs, if sufficient benefits can be identified, that particular type can still be selected and costs related to acquiring the equipment has to be accounted for.

In case the design loads on the rib are relatively low, one may select the rib type that is manufactured by forming sheet or plate material into a rib. The additional stiffnessmay be added by adding stiffening profiles to that rib.

In case the loads are relatively high, i.e. near landing gears, a more rigid and stiff design is required. This may be achieved by forging and machining a rib from thick plate material.

A distinct difference between the two rib types is the difference near radii. In case of forming plate material, the minimum bending radius is dictated by the properties of the material. Smaller radii will lead to failure during bending. In case of machining, similar inner radii with smaller outer radii can be achieved by milling, which provide a (geometrically) stiffer connection with the surrounding structure.

5.5.2 Spars

The main function of the spars in a wing structure is to carry the wing bending loads.As explained in section 5.3.2, the spars are beams designed to carry bending by girders carrying the normal forces and web plates that carry the shear forces, see also Figure 5.21.

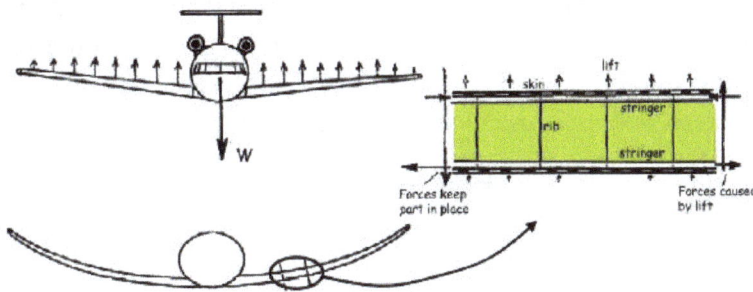

Figure 5.21: Forces in the wing structure as result of upward wing bending. (TU Delft, n.d., 5-21.jpg. Own Work.)

The upward bending illustrated in the lower left part of Figure 5.21, requires some additional attention. Zooming in a bit closer, it is evident that the elements in the wing deform as illustrated in Figure 5.22. To carry the bending loads and to resist against this deformation, the diagonal elements are very important. And as explained before, the diagonal function can also be carried by sheet material, or web plates in this matter. As a result, the basic form of spars is often the I-beam, in which the girders pick up the normal forces and the web plate the shear forces.

Although the deflection near the wing root is significant smaller than near the wing tip, the loads near the wing root are significantly larger. The equilibrium between the normal forces in both girders of the spar near the wing tip and the upward bending moment, results in relatively small forces. Near the wing root, equilibrium must be provided with the upward bending moment of the complete wing. This results in veryhigh forces near the wing root.

As a result the spar, the girders, and web plate, must be thicker near the wing root than near the wing tip. However, extrusion of an I-beam (see chapter 4 for the extrusion process) can only be performed on constant cross-sections. A spar with varying cross-section along the wing span cannot be extruded.

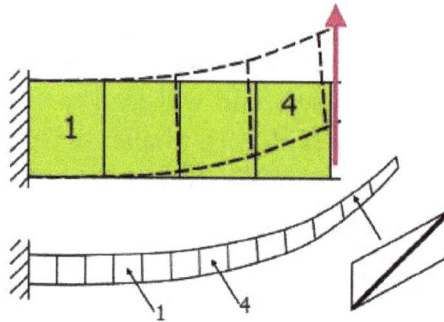

Figure 5.22: Deformation related to the upward wing bending. (TU Delft, n.d., 5-22.jpg. Own Work.)

Design concepts to deal with this manufacturing aspect are based on building up the spar from different elements. Either the girders are separately manufactured andconnected to a web plate by riveting for example, or a spar with constant thickness is extruded and subsequently reinforced by bonding additional sheet material to webplate and girders. This is illustrated in Figure 5.23.

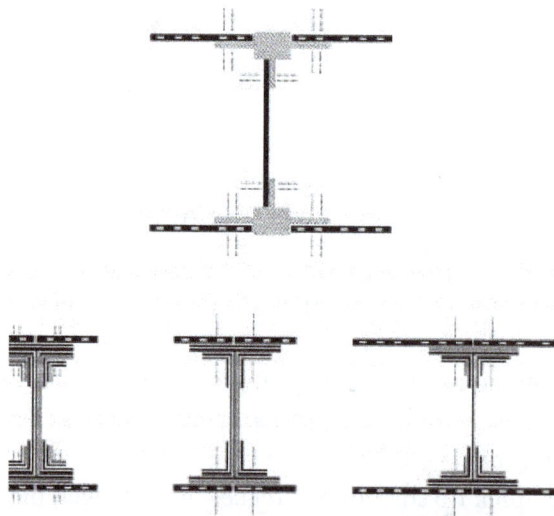

Figure 5.23: Build-up concepts to increase the cross sectional area of the spar towards the wing root. (TU Delft, 2018, 5-23.jpg. Own Work.)

Similar to the discussion of the wing ribs, it can be concluded that depending on the function of particular spars, the loads it has to carry, the design philosophy and the available manufacturing equipment, different spar types can be selected, see Figure 5.24.

This also implies that the spar caps (girders) can be manufactured by extrusion and or machining, or by forming sheet material, see Figure 5.25. The forming concept has less geometrical stiffness compared to the extruded concept.

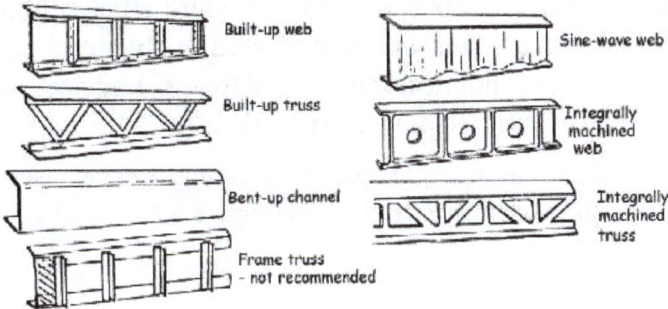

Figure 5.24: Illustration of the different spar types. (TU Delft, N.D.)

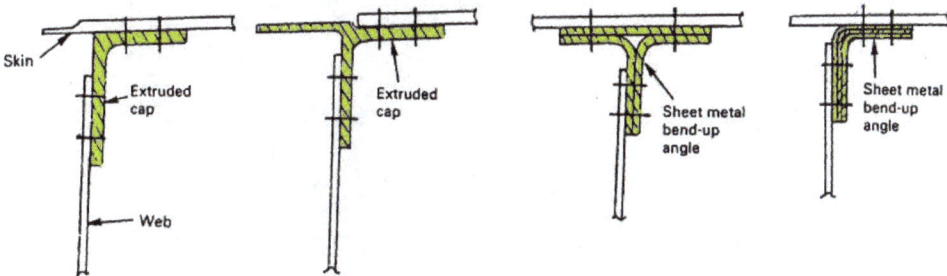

Figure 5.25: Illustration of the spar cap types; extrusion (left) and forming. (TU Delft, N.D., (right)

5.6 TORSION BOX

The wing is in general loaded by bending and torsional moments. It has been explained that the bending moment can be easily carried by the spars. However, if a wing contains only a single spar, it will not be able to carry the torsional moment, because an I-beam has low resistance against torsion.

In case the wing is equipped with two spars, torsional moments can be carried by differential bending. Differential bending means that torsional deformation is translated to an upward bending of the rear spar and a downward bending of the front spar. In this case, the spar is carrying bending loads for which it is designed.

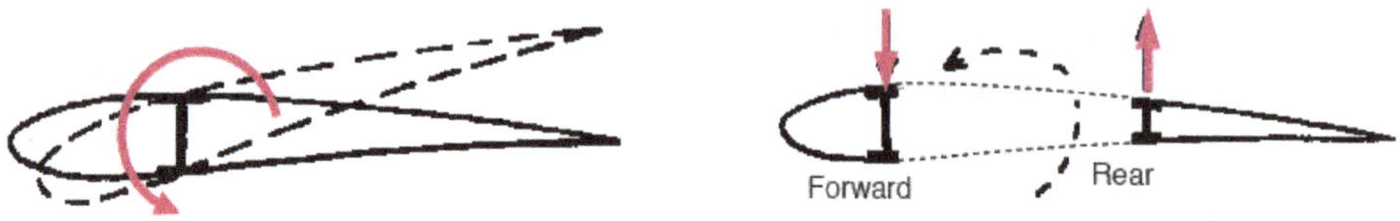

Figure 5.26: Torsional moment in case of single (left) and two spars (right). (Alderliesten, 2011)

However, despite that the resistance to the differential bending is well provided by the spars, torsional moments can be carried more efficiently by a cylinder or closed box, see Figure 5.27. This will be discussed in chapter 6 in more detail. In fact, the cross section does not necessarily need to be cylindrical; any cross section may do, as long as it remains a closed cross section.

A historical example of a single spar wing structure that uses the cylindrical torsion tube is the Blackburn Duncanson, see Figure 5.27.

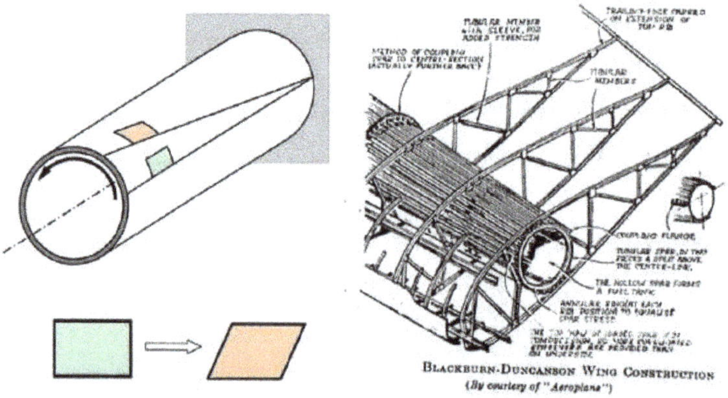

Figure 5.27: Resistance to a torsional moment by a cylinder (left) and the Blackburn Duncanson single cylindrical spar example (right). Derivative from left: Alderliesten, 2011, 5-27-a.jpg, Own Work; and Aeroplane, N.D.)

The torsional resistance of a closed box structure can be exploited in a wing structure, especially, because the cross section does not need to be circular. This is illustrated in Figure 5.28. Compared to the two spar differential bending concept, the torsion box contains thicker skins. The functions of these skins are to take up the aerodynamic forces and to contribute to the torsion box. In that case they partially take over the bending function of spar caps. As a consequence the spar caps may have smaller cross sections than in case of the two spar concept. The web plates of the spar will be thicker similar to the wing skins; they contribute to the torsion box and add to the bending resistance.

The stringers in this concept reinforce the upper and lower wing skin and by doing so, also partially contribute to the function the spar caps have in the two spar concept.

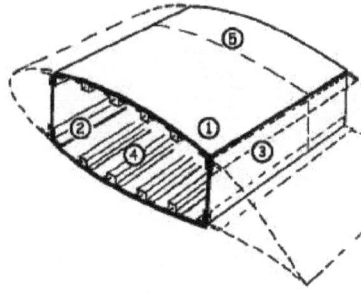

Figure 5.28: Illustration of the different structural elements in a torsion box. (TU Delft, N.D.)

The advantages of the torsion box over the two spar concept is that a completely unsupported and load bearing structure can be achieved that does not require support or struts. At given wing span, the wing can be thinner in case a torsion box concept is applied, or similarly, at given wing thickness the wing can be longer. Additionally, by designing the torsion box carefully, the torsion stiffness and the bending stiffness can be engineered separately.

In general, the torsion box concept results in lower structural weight compared to the two spar concept.

5.7 STRUCTURAL DETAILS

Considering the fuselage and wing structures, the elements and their functions have been discussed. However, by creating a three-dimensional structure one will realise that the intersections between the different structural elements require additional considerations. Four examples are discussed here; stringer joggling, stringer couplings at the location where two fuselage barrels are joined, stringer frame intersections, and stringer rib intersections.

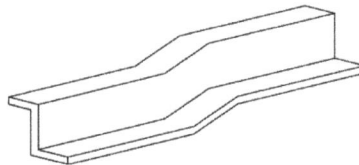

Figure 5.29: Illustration of stringer joggling. (Alderliesten, 2011)

Because the loads are not constant throughout the fuselage structure, the required thickness of fuselage skin panels will vary. The outer surface of the fuselage must be undisturbed for aerodynamic reasons, which implies that thickness steps in the fuselage skin are provided at the inner side of the panel. This can be achieved by either milling thickness step out of thicker sheet material, or by adding additional layers, or so called doublers. These doublers can be riveted or bonded to the skin structure.

The stringers that are attached to the skin have to follow these thickness steps. A common method is to joggle the stringers. Joggling of a stringer implies a shear deformation of the stringer, as illustrated in Figure 5.29.

A fuselage structure is built up from different panels that are connected by longitudinal and circumferential joints. This implies that the panels are manufactured including stringers, but that once the panels are joined in assembly, the stringers can be connected to each other. This is illustrated in Figure 5.30. The joint between the fuselage skin panels is created by adding splice plates at the inner side of the structure, to obtain an aerodynamically smooth surface outside. These splice plates also cause thickness steps that have to be accounted for by the stringers that are attached to the fuselage skin. When a stringer coupling is used to connect the stringers of both panels, the design of the coupling needs to account for the thickness step.

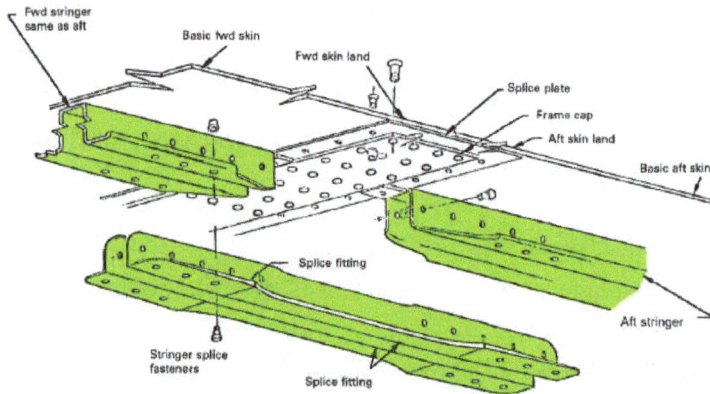

Figure 5.30: Illustration of stringer connection at the joint between two fuselage barrels. (TU Delft, N.D.)

Another intersection between stiffening elements that has to be considered is the intersection between stringer and frame. To create the three-dimensional geometry here, there are in general two concepts to provide the intersection. In both cases, the stringers are continuous and not interrupted to have the optimal longitudinal stiffness of the panel.

In the first concept, the frame is locally interrupted to provide an opening for the stringer. This method implies a local weakening of the frame and reduction in bending resistance of the beam. The other concept positions the frame above (floating) the stringers and provides the connection between frame and fuselage skin by castellation or clips. An illustration of both concepts is given in Figure 5.31.

Similarly to the frame stringer intersection, intersections between stringers and ribs have to be created in wing structures. Figure 5.32 illustrates that for this case there are in general three possible solutions to enable the intersection;

- ◉ Both the rib and stringer are uninterrupted
- ◉ The stringer is interrupted
- ◉ The rib is interrupted

The selection of the most appropriate concept depends on the loads that act locally upon the structure and the manufacturability of the concept.

Figure 5.31: Illustration of intersections between frames and stringers in a fuselage panel. (Kolossos, 2006, CC-BY-SA 3.0)

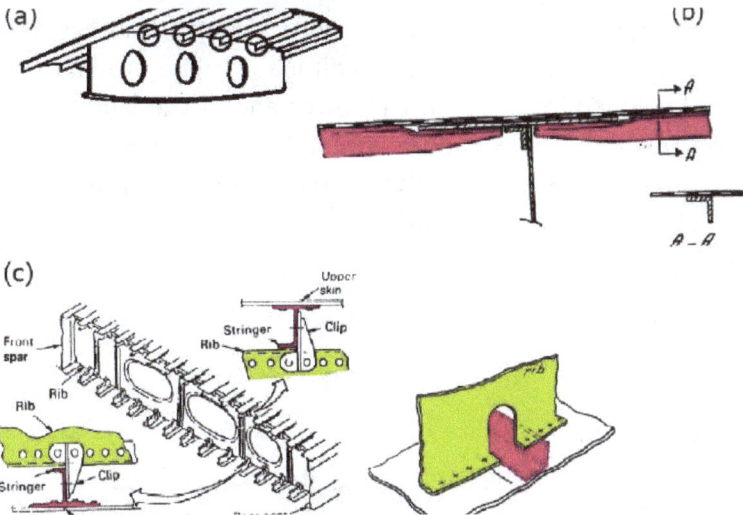

Figure 5.32: Illustration possible rib stringer intersections; both uninterrupted (a), stringer interrupted (b), rib interrupted (c) shows both possibilities: inserts are both uninterrupted, using the clips. The close-up shown on the right is rib interrupted. (TU Delft, N.D.)

5.8 TYPICAL SPACECRAFT STRUCTURES

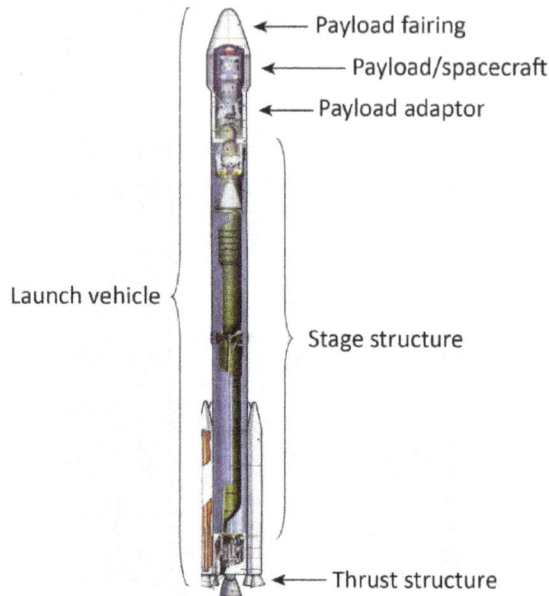

Figure 5.33: Illustration of the spacecraft and launch vehicle (Delta II), with typical components. Derivative from: NASA (2000), Public Domain.

When looking at the spacecraft, one has to take into account that space despite being the operational environment of the spacecraft, does not solely determine the structural design. In fact, the main structural items that can be observed in the spacecraft are dimensioned to sustain all loads related to launch. This means that the forces provided by thrust at the lower end of the launch vehicle have to be transferred through the structure, including the spacecraft structure. To accomplish this load transfer through the spacecraft structure, two categories of spacecraft structures can be distinguished

- Strutted structures
- Central cylindrical shell structures

Both types of structures will be briefly discussed in this section.

5.8.1 Strutted structure

Part of the launch vehicle is the payload adaptor that is discussed in section 5.9.4. One may assume that the load provided by thrust to be transferred through the spacecraft structure is inserted in to the structure by the adaptor.

One way to create the necessary load paths through the structure is by using struts. These struts basically are a form of truss structure, discussed in section 5.2.1, where the struts are loaded primarily in compression in their longitudinal axis. An example of a truss structure is given in Figure 5.34.

Figure 5.34: Example of a spacecraft utilizing a truss structure. (ESA, 2012, Copyright ESA, Use Permitted.)

An important dimensioning aspect then will be the stability of the trusses, see chapter 8. The trusses should not bulge out in compression, but remain stable to be able to transfer the load to the remaining structure. However, because the primary function of the trusses is to sustain this particular load case, they can be optimized for this case with particular material and shape solutions, like, for example, the struts in the Space Shuttle ribs mentioned in section 5.2.1.

5.8 .1 Central cylindrical shell structure

The central cylindrical shell structure configuration distinguishes itself from other structures, because it has a major central thrust-load-bearing member in the form of a cone of a cylinder (Wertz & Larson, 1999). All load provided by thrust and inserted via the adaptor is transferred through a central cylinder that is primarily dimensionedfor this load case.

All systems within the spacecraft are attached at strong points directly to the cylinder or they are attached by combinations of struts, platforms and shear webs. An example is illustrated in Figure 5.35.

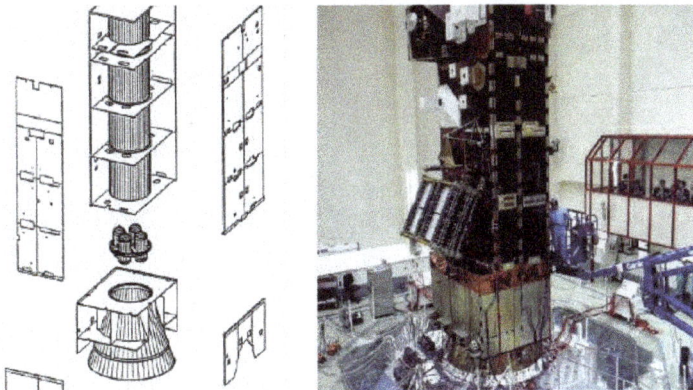

Figure 5.35: Example of a (polar platform) spacecraft utilizing a central cylindrical shell structure. Derivative from ESA, (2002), Copyright ESA, Use Permitted.

5.9 TYPICAL LAUNCH VEHICLE STRUCTURES

In this section the four main structural parts of the launch vehicle are discussed: the payload fairing, the stage structures, the thrust structures and the adaptors.

5.9.1 Payload fairings

The payload fairing is an important part of the launch vehicle. It protects the payload during launch and ascent against the aggressive environment, i.e. aerodynamic heating and pressure. Once the launch vehicle is outside the atmosphere, the fairings are disposed, exposing the payload to the environment. The consequence of the fairing disposal is that the to be accelerated mass will decrease, which increases theefficiency of the thrust.

5.9.2 Stage structures

Rocket launchers typically consist of multiple stages, which each contain a rocket propulsion system with several subsystems within the structure. These stages can either be placed in sequence, i.e. on top of each other, or in parallel. The main stage forms the core of the launcher, while booster stages are often placed in parallel to increase the thrust during the initial part of the ascent. The booster stages are illustrated in Figure 5.33.

Different design concepts could be chosen for the stage structure. The two most important design concepts can be characterized by whether the fuel tanks are part of the load bearing structure, or not.

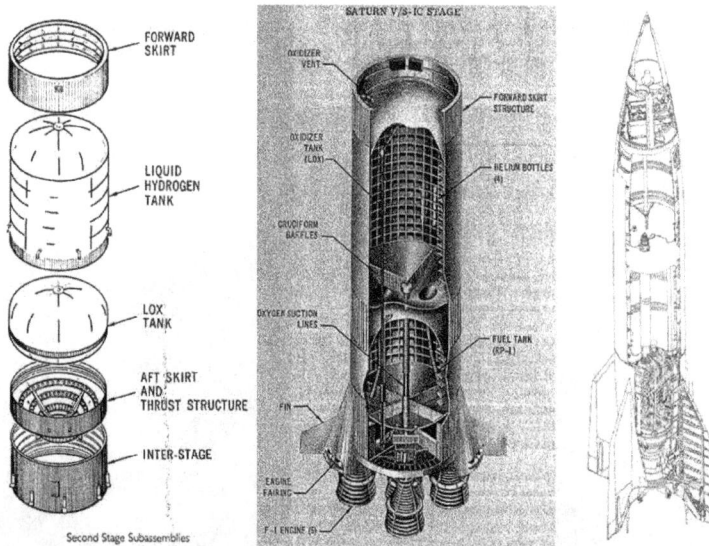

Figure 5.36: Illustrations of typical stage structures: Saturn, fuel tank part of the structure (left) and V2, fuel tank not part of the structure (right). Derivative from NASA, (n.d.), Public Domain.

The exploded view in Figure 5.36 illustrates the assembly of the structural components in case the fuel tanks are part of the load bearing structure. The structural components are connected with interface structures to transfer the load from one component to another. The skin concept is usually referred to as semi- monocoque, i.e. load bearing skin containing internal stiffeners, as explained earlier.

The design concept in which the fuel tanks are not part of the primary structure contains an external skin that is separated from the internal tanks by longerons and circular stiffeners, this is illustrated in the right part of Figure 5.36

5.9.3 Thrust structures

The thrust structure is an important part of the launcher as all loads introduced by the propulsion system is transferred via this structure into the main stages of the launcher. The high loads (Ariane 5 about 15000 kN at take-off ~ 1500 tons) are introduced rather concentrated and must be distributed into the stage structure via aconical structure, as illustrated in Figure 5.37.

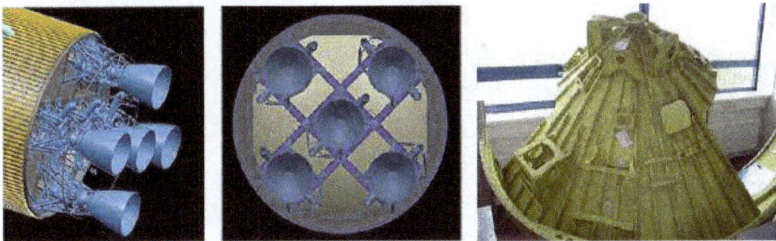

Figure 5.37: Illustrations of ARES V thrust structures (left -NASA, 2007, Public Domain) and section of the Ariane 5 HM60 conical thrust structure (engine mounted on top of cone) (right – TU Delft, N.D.)

5.9.4 Adaptors

A similar function of transferring loads, as mentioned for the conical structure that ispart of the thrust structure, has to be fulfilled by the adaptors. This part is located in the upper part of the structure and has the primary function to transfer the loads from the stages into the spacecraft or satellite located in the payload area under the fairing.

Hence similar conical shapes can be observed, as illustrated in Figure 5.38, where instead of the load distributed to a wider area, the loads are rather concentrated into the primary structure of the spacecraft, especially when the spacecraft applies acentral cylindrical shell design, see section 5.8.2.

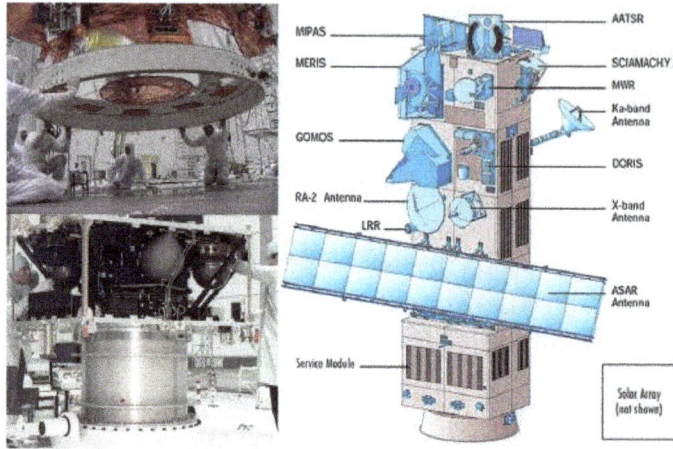

Figure 5.38: Illustrations of adaptors (lower left image shows the Hipparcos main structure and fuel tanks, upper left image shows the adaptor for the Envisat shown right). Derivative from ESA, (2002), Copyright ESA.

CHAPTER-6
AIRCRAFT & SPACECRAFT LOADS

6.1 INTRODUCTION

In the previous chapter, typical aircraft and spacecraft structures have been discussed. The structural elements belonging to the airframe fulfil primarily load bearing functions. After reviewing the structural elements and their characteristics, it is time to consider the loads acting on these structures and elements.

This chapter will review typical loads and load cases for both aircraft and spacecraft. These loads can be divided into concentrated and distributed loads, which each can be divided into static or dynamic loads. These characteristics together with the typical load paths will be discussed in this chapter.

Figure 6.1: The four principle forces considered acting on an aircraft in flight. Derivative from Mr_worker, 2014, Public Domain.

6.2 EXTERNALLY LOADED AIRFRAME

To assess the flight performance, an aircraft in horizontal steady flight is typically illustrated with the principle forces that act on such aircraft; lift in equilibrium with weight, and thrust in equilibrium with drag, see for example Figure 6.1. These equilibriums change for various manoeuvres, but are commonly based on the major forces.

However, although sufficient for representing the forces and equilibriums when describing aircraft performance, this presentation becomes insufficient when considering structures and loads acting on those structures. For example, the presentation of Figure 6.1 may be more accurately presented by the illustration in Figure 6.2.

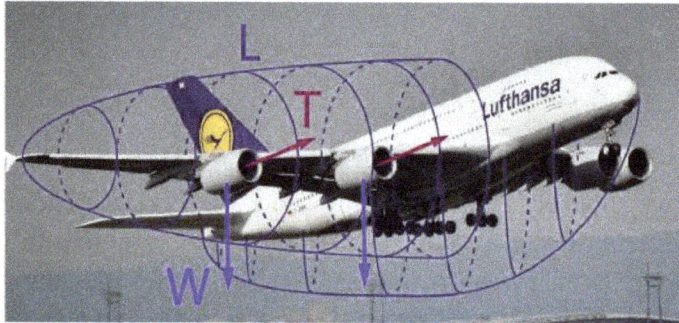

Figure 6.2: Concentrated and distributed forces acting on an aircraft in flight. Derivative from Mr_worker, (2014), Public Domain.

The presentation of Figure 6.1 for instance, provides equilibrium for the complete aircraft, but it does not provide a realistic presentation for the wing structure. For the wing, Figure 6.1 is insufficient as it does not show the forces acting on the wing. In fact, it assumes (equilibrium of) all forces are acting on the centre of gravity. Here, a presentation like in Figure 6.2 seems to give a better picture; the lift generated by the wing is partly compensated for by the weight of the wing and engines, while the thrust is to an even lesser extent compensated by the drag of the wing. The resultants are of course acting on the centre of gravity creating equilibrium with the fuselage and empennage of the aircraft.

In general, the airframe is loaded by a combination of loads, or load cases, which may be attributed to aircraft manoeuvres, gusts, cabin pressure, landing, etc. The loads act on the structure either by concentrated or distributed forces. For example, the thrust is illustrated in Figure 6.2 as a concentrated load, while the wing lift is represented as distributed force acting on the wing. In fact, the engines apply the thrust to the wing via few bolts (concentrated loads), while the lift is a force resulting from pressure multiplied with the area it acts on (distributed force). The consequence of concentrated or distributed loading acting on a structure is illustrated with the example of a beam in bending.

Example: Bending of a beam

Consider the beam illustrated in Figure 6.3 rigidly clamped at one side and loaded at the other side by a concentrated load at the end of the beam or a distributed load decreasing towards the tip of the beam. If both load cases represent the same total force P in vertical direction, then the deflection of the beam for the concentrated and distributed case can be given by

$$\delta = \frac{PL^3}{3EI} \text{ and } \delta = \frac{Pl^3}{15EI}$$

respectively, where P is the total vertical load, the length of the beam, E the Young's modulus and I the area moment of inertia.

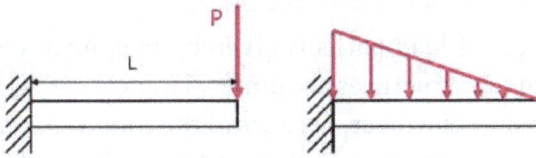

Figure 6.3: Beam bending induced by a concentrated force (left) and a distributed force (right). (Alderliesten, 2011)

6.3 LOAD PATH

To understand the meaning of loads acting on a given structure, first the load path must be understood. A load path is a physical trajectory that links the location of applied force and forces elsewhere that provide equilibrium with the applied force. However, this trajectory should be able to carry and transfer the loads. This isillustrated with the point force acting at a certain location in space, as illustrated in Figure 6.4

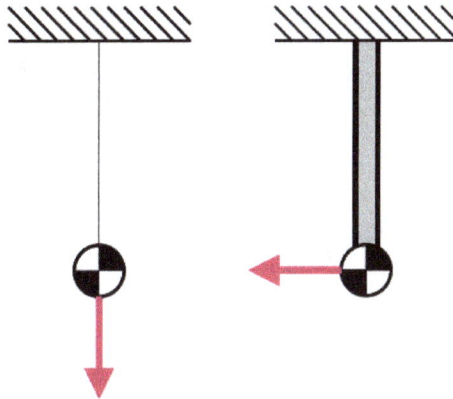

Figure 6.4: Different load paths for a point force acting in space. (Alderliesten, 2011)

For the case where the force is applied downward in vertical direction, there are several solutions to provide equilibrium. A simple cable or rod may be sufficient to be connected between the point force and the wall. In this case the cable or rod represents the load path and it is loaded in tension. For the second case in Figure 6.4, however, the cable may not be sufficient, because the point force would cause the cable to rotate. As a result, equilibrium would not be obtained and thus the cable cannot provide the load path.

The solution in that case may be to connect the point force with a beam to the wall. The beam would resist bending and equilibrium would be obtained not only with a horizontal force at the wall in opposite direction of the applied point force, but as will be explained later in this chapter, also by a set of forces in vertical direction.

6.3.1 Cathedral structures

An illustrative example of load paths is given by looking at old cathedrals. These thin-walled and often tall structures are dimensioned to carry the weight of its own structure in compression. However, because of the thin walls and the tall structure, the addition of the roof structure on top of the walls would result in resultant forces pushing the walls away from each other. Because the walls can only carry load in the plane of its structure, this would lead to collapse of the building. There are two basic methods to avoid this type of failure (see Figure 6.5)

- ◉ Using tensile elements to provide equilibrium in the roof structure (inside)

- ◉ Using buttresses to provide a load path through the cathedral structure (outside)

Figure 6.5: Two concepts to sustain the loads from the roof structure's weight. (Alderliesten, 2011)

Because the materials and corresponding production techniques were not yet sufficiently developed to manufacture elements that could be loaded in tension, the designers of these structures used buttresses. The design of buttresses is based on the concept that the resultant force induced by the roof structure can be directed further downward by adding additional forces directed downward. These additional forces are applied by the weight of the buttress structure.

Comparing the different churches and cathedrals, one may observe differences in the design of the buttresses. The weight of the buttresses necessary to provide a resultant force path to the ground can be obtained with three solutions:

- ⊙ Creating wide buttresses or long flying buttresses to achieve sufficient distance from the central structure to be in line with the load path for the lateral forces

- ⊙ Creating very thick buttresses to create sufficient weight to provide a resultantload path within the buttress

- ⊙ Creating very tall buttresses with long decorated peaks of which the weight provides a resultant load path through the relatively thin buttresses.

Figure 6.6: Illustration of a central cathedral structure and buttresses to support that structure. Derivative from Roddy, (2007), UC Davis.

6.3.2 Bending of beam structures

Consider the deflection of a beam as illustrated in Figure 6.7. The vertical load applied at the free end of the beam will cause a deflection. Looking at the left hand side illustration in that figure, the load path obviously consists of normal forces at the upper and lower surfaces (or girders) and shear forces in the web plate, see section 5.3.2.

For both cases (a) and (c) in Figure 6.7, the load path is in equilibrium with the applied force in the centre of the cross section. However, for case (b) the shear forces in the web plate are not aligned with the applied force. The distance between the centre of the web plate, see Figure 6.8 (a), and the centre of the cross section will induce a resulting moment. This moment is in fact a torsional moment, which will rotate the beam together with the vertical deflection.

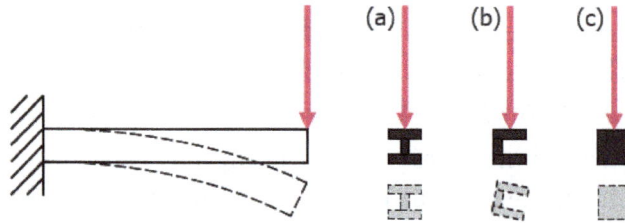

Figure 6.7: Bending of a beam with different cross sections. (Alderliesten, 2011).

Concerning the load in the elements of the beam, it has been mentioned before that the forces at the clamped side of the beam are high. The magnitude of these forces depends on the applied load at the other end and the length L of the beam. The longer the beam, the higher the normal forces in the girders at the root.

This equilibrium can be illustrated with Figure 6.8 (b); the applied load multiplied with the length L must be in equilibrium with the product of normal forces and height H. This means that at the end of the beam the normal forces are zero and they linearly build up towards the root. Taking a section of the girder at any location of the beam, as indicated by the square in Figure 6.8 (b), and looking at its force equilibrium, it can be observed that equilibrium is provided by the increase of normal forces and the shear forces that the web plate applies to the girder.

Here, it is evident that the shear forces in the web are constant throughout the web plate. Cutting away a part of the web plate, see Figure 6.8 (c), means that the normal forces in the girders are not increasing in the area where the web plate is removed. Only due to the constant shear forces that the web plate applies to the girders, the normal forces increase. Without the shear forces, the normal forces would be constant.

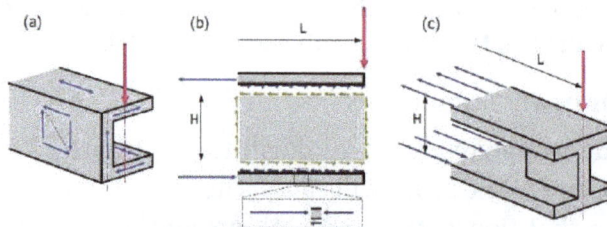

Figure 6.8: Specific aspects related to beam bending (Alderliesten, 201)

6.4 LOADS AND LOAD PATHS IN AN AIRFRAME

It has been discussed a few times, that due to the lift generated by the wing, the wing will bend upward. All lift generated by the wings will be in equilibrium with the weight of the wing and the fuselage. But if the aircraft is considered as two beams that cross, then the lift will provide an upward bending of the wing beam, while the weight of the fuselage beam will cause a downward bending over the wing. This deformation is illustrated in Figure 6.9. Both the lift generated by the wings and the weight of wing and fuselage are considered to be distributed forces acting on the structure.

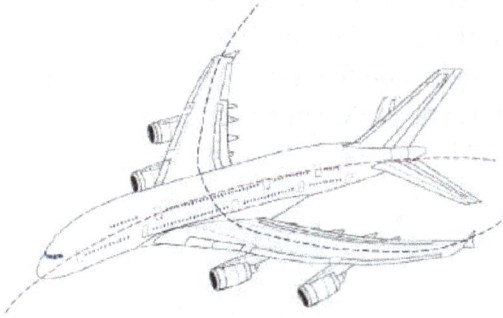

Figure 6.9: Upward wing bending and downward fuselage bending. (Alderliesten, 2011)

Examples of concentrated loads are the forces applied to the aircraft by the landing gears, see Figure 6.10. If the aircraft stands on the platform, the forces introduced by the landing gears are in equilibrium with the total weight of the aircraft.

Figure 6.10: Concentrated forces applied to the airframe by the landing gears. Derivative from Lampel, (2013), CC-BY-SA3.0.

The load path that can be identified in this case starts at the local load introduction of the landing gear, via the rib at that location into the wing structure. Here, the normal forces are carried via the skins, while the shear forces are carried by the web plate of the spars. Although Figure 6.10 only illustrates the load path from the main landing gear towards the fuselage, also part of the load should provide equilibrium with the weight of the wing (and engines) towards the wing tip.

Let's consider the aircraft as it is standing on the platform. Wind may apply a load on the vertical tail plane as illustrated in Figure 6.11. If the vertical tail plane is considered to be clamped at the fuselage, the deformation may be represented by a beam under bending, as illustrated in the right hand side of the figure. Notice that as the wind applies to the complete surface of the vertical tail plane is considered a distributed load. Because the frontal surface decreases towards the tail tip, the forces most likely will reduce towards the tip.

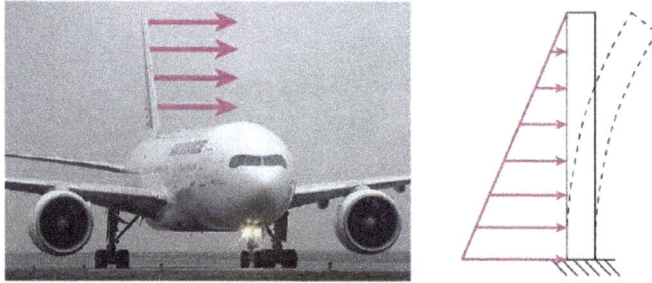

Figure 6.11: Bending moment on vertical tail plane induced by wind. Derivative from Noret, (2012), CC-BY-SA2.5.

Thus the forces are in equilibrium by forces at the root of the vertical tail. However, if the fuselage is considered to be a cylinder which is 'clamped' at the location of the wing, then the bending moment on the vertical tail will rotate the fuselage, as illustrated in Figure 6.12.

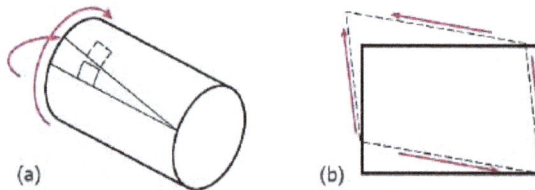

(a)   (b)

Figure 6.12: Rotation of the fuselage (a) with as result shear deformation of the skin panels (b). (Alderliesten, 2011)

Zooming in further onto the fuselage structure, the fuselage skin panels will deform in shear due to the torsion of the fuselage.

The fuselage was assumed to be 'clamped' in the area of the wing. If the aircraft is standing on the platform, the bending moment on the vertical tail and torsional moment on the fuselage will be in equilibrium with the main landing gear.

The reaction forces introduced by the landing gear will be different in magnitude, where thedifference between these forces will be providing the equilibrium.

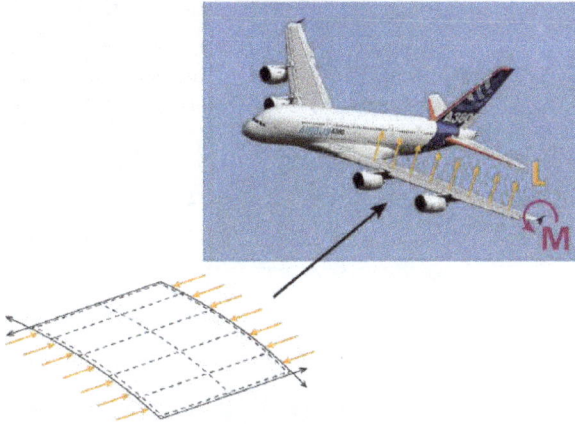

Figure 6.13: Upward bending and rotation of the wing. Derivative from: Ienac, (2006), Public Domain.

A similar discussion may apply to the bending and rotation of the wing in flight. First,the wing may be considered clamped at the root and bending is carried by the upper and lower wing skins in respectively compression and tension forces. The torsional loading will be carried by shear in the torsion box, i.e. both wing skins and both web plates of the spars.

In cruise, the loads will be similar for both wings, which imply that the forces acting on the centre wing box in the fuselage will be equal in magnitude but opposite in sign.

Here, the example of the cathedral structure may explain more specifically how load should be introduced into a structure. Consider the A400M shown in Figure 6.14. The connection between the wing structure and the fuselage structure could be designedin different ways. Two possible solutions are illustrated in Figure 6.15.

Figure 6.14: Lockheed C-5 Galaxy transport aircraft in maintenance (left) and the centre wing box of a Lockheed C-130 Hercules transport aircraft (right). Derivative of Sapp – U.S. Air Force, (2006, 2011), Public Domain.

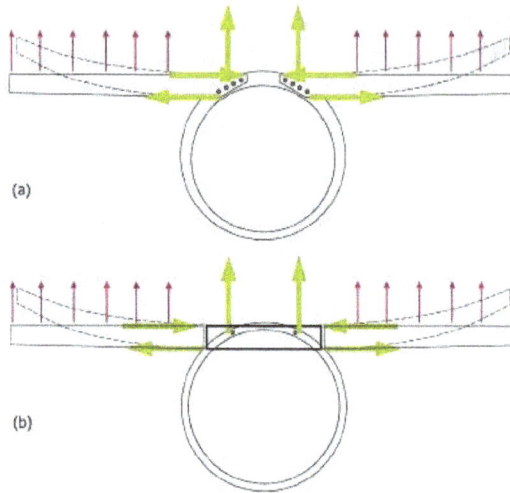

Figure 6.15: Illustration of two solutions to connect the wing structure to the fuselage structure. (Alderliesten, 2011)

The first solution, Figure 6.15 (a), is not preferred, because it will not only introduce the lift generated by the wing into the fuselage structure, but also all bending deformation as result of that lift. Locally the frames of the fuselage will be highly loaded by tension and torsional forces, which can only be achieved by designing very large and thick frames. But even then, it may be disputable whether the design would suffice.

The second solution, Figure 6.15 (b), is the preferred solution and in fact the solution applied on the A400M and other aircraft. The centre wing box is designed such that the normal forces introduced by the upper and lower wing skin due to wing bending are carried by the upper and lower skin of the centre wing box. Only the lift is introduced into the fuselage structure as vertical force by four attachment points (lugs).

In that respect, this example may be compared to the example of the cathedral structure. The roof due to its weight introduces forces that push the walls away from each other. The solution applied in the past was to guide all these forces through the structure towards the ground using buttresses.

However, a leaner solution resulting in significant less weight would be to provide equilibrium inside the roof for the forces acting outward, and to guide only the vertical component of the weight into the remaining structure, see the left hand side of Figure 6.5.

6.5 COMPLEX LOAD CASES

In this section two more complicated load cases for aircraft structures will be discussed: fuselage loads and dynamic loads.

6.5.1 Fuselage loads

In general, one has to be aware that the many possible load cases acting on the structure either as concentrated loads, or as distributed loads, provide a complex picture of areas on the structure that are dictated by different load cases and damage scenarios.

This is illustrated in Figure 6.16. Static ground loads can be significant near the landing gears, but will be less restrictive in other areas. The earlier mentioned downward bending of the fuselage over the wing due to its weight will imply longitudinal tension in the upper part of the fuselage. Together with the circumferential tension as a result of pressurization (which also adds to the longitudinal tension, see next chapter), this will make tension and fatigue dominant

load cases for the upper fuselage. The lower fuselage however, will be predominantly facing compression loads due to the fuselage bending, with as consequence dominant criteria on stability.

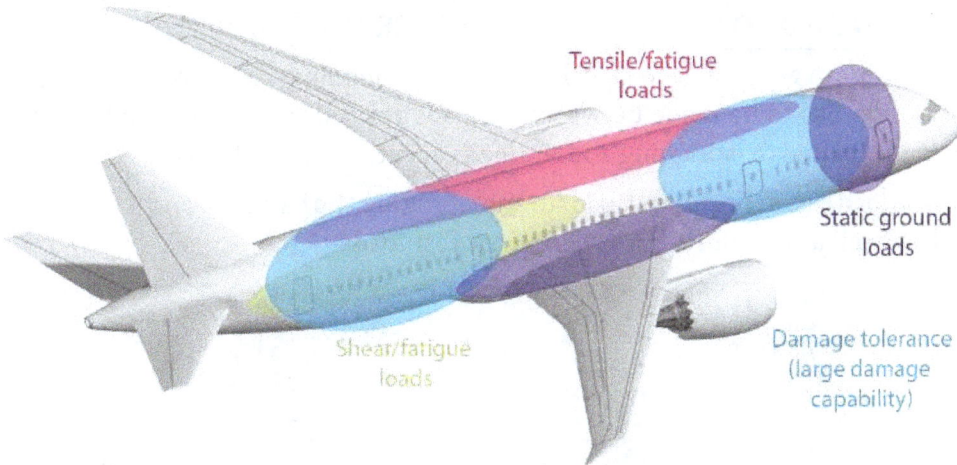

Figure 6.16: Illustration of the dominant load cases and design criteria in the fuselage structure. (TU Delft, 2018)

The aft side fuselage shells will face shear and fatigue loads, introduced by the loadsintroduced by the empennage. One example of the bending of the vertical tail plane has been discussed at the beginning of this section.

6.5.2 Dynamic loads

The discussion of loads and load paths in this chapter has been limited implicitly to quasi-static loading, i.e. an external load is applied to the structure and equilibrium must be provided. However, in an aircraft not all loads are quasi-static. Some loading appears to be very dynamic of nature. Although fatigue, discussed in chapter 10, relates to cyclic loading, it relates in most cases still to quasi-static loading. For example, the pressurization cycle (equivalent to a single flight) may be analyzed as a single load cycle in an almost constant fatigue loading spectrum over the year. Nonetheless, the load cycle itself is considered quasi-static.

Sometimes interaction of aerodynamic forces, structural elastic reactions, and inertia of components may induce oscillations of aircraft components. These oscillations may be quite significant and severe. One example of these oscillations may be flutter, which may lead to destruction of components if not anticipated for.

Another example is buffeting caused by airflow separation from one component onto another. A well known example here is the buffeting issues related to the two verticaltails of the F18 Hornet. Because the oscillations may increase if natural frequencies of the structure are approached, the design has to take into account the damping properties of the structure or component. In addition, the design should account for natural frequencies and attempt to keep the natural frequencies outside the range of potential frequencies of oscillations. An example of this approach is discussed for spacecraft structures in chapter 8.

6.6 LOAD AND LOAD CASES FOR SPACECRAFT STRUCTURES

Spacecraft in general face different types of loading compared to aircraft. A distinction has to be made between the loads related to handling and pretesting, the loads related to launching and the loads occurring once the structure is in orbit.

The load cases considered prior to launching are the handling and transportation loads as well as the vibration and acoustic test loads.

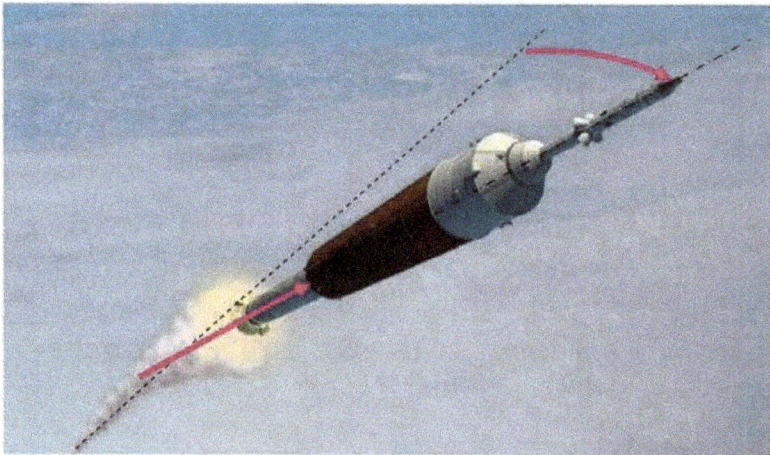

Figure 6.17: Illustration of the axial and lateral loads and movements. Derivative of NASA (2006), Public Domain.

The static and dynamic loads during launching can be categorized by:

⦿ Quasi-static

- ⊙ Sine vibration

- ⊙ Acoustic noise and random vibration

- ⊙ Shock loads

The consequence of these loads for design is discussed in detail in chapter 8. In orbit the spacecraft will experience shock loads, structurally transmitted loads, internal pressure and thermal stress. These are generally much lower than launch loads and are not discussed in detail in this book.

The steady state loads can be directed either in axial direction of the launching vehicle or perpendicular to that direction (lateral direction). The main load case acting in axial direction is the thrust generated by the engine of the launching vehicle. Wind gust and vehicle manoeuvres are generally oriented in lateral direction.

One important distinction one has to make here is the primary axis of the spacecraft (satellite, etc) with respect to the primary axis of the launching vehicle. The explanation of axial loads and lateral loads may seem obvious for the case illustrated in Figure 6.17, but when designing a spacecraft, it depends on the orientation of the spacecraft with respect to the launching vehicle, see Figure 6.18.

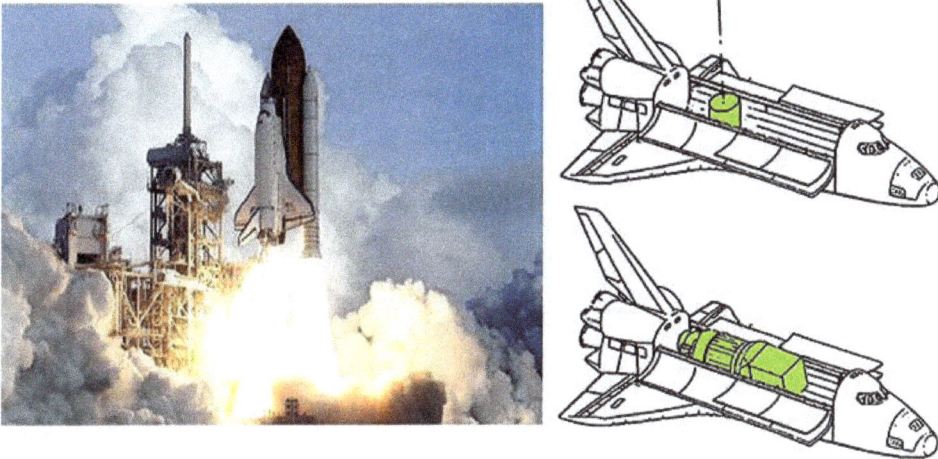

Figure 6.18: Illustration of spacecraft orientation with respect to launching vehicle. Derivative of NASA, (2000), Public Domain.

The upper case in the right hand side of Figure 6.18 has its primary axis perpendicular to the maximum acceleration, whereas the lower case has its primary axis parallel to the maximum acceleration. The design of the spacecraft will be different for both scenarios.

CHAPTER-7

TRANSLATING LOADS TO STRESSES

7.1 INTRODUCTION

In the previous chapter, the loads acting on an aircraft or spacecraft structure have been discussed. These loads are generally considered to act at certain locations on the structure either by concentrated loads or by distributed loads. These loads induce deformations in the structure that correlate to stresses within the structure. To understand this, one should reconsider the example of the tensile test in chapter 1, where the load acting on the specimen induces an elongation (deformation) of the specimen, which corresponds to the stress in the specimen, calculated with the load divided by the cross section.

In this chapter, some cases are dealt with to explain how loads acting on an aircraft or spacecraft structure will induce deformation, but also lead to stresses within the structure. These stresses must be known, to be able to assess whether the materials used in the structure are capable to sustain the loads. Here, the stresses calculated from the loads acting on the structure should be compared with the stress-strain response of the materials to identify whether sufficient strain is present, and whether the material's stiffness will be sufficient to limit deformations to acceptable levels.

7.2 PRESSURIZATION OF A FUSELAGE STRUCTURE

As commercial aircraft commonly operate at altitudes of 10000 meters, aircraft fuselages are pressurised to ensure that crew and passengers can breath normally. In this section the stresses induced into the structure as a result are explained.

7.2.1 Stresses in cylindrical pressure vessel

The circumferential and longitudinal stresses in a pressure vessel can be derived from equilibrium within the vessel. To calculate the circumferential stress, one first considers the upper half of a pressure vessel with unit length as illustrated in Figure 7.1.

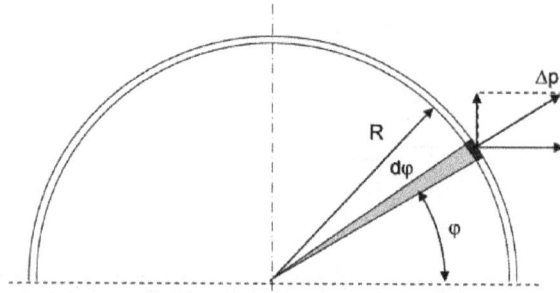

Figure 7.1: Equilibrium between radial pressure distribution and circumferential stress per unit vessel length. (TU Delft, N.D.)

There should be equilibrium in vertical direction between the circumferential stress and the vertical component of the pressure p acting at the inner surface of the σ_{circ} vessel. Per unit length of the vessel, this equilibrium can be described with

$$2F_{circ} = 2\sigma_{circ}t = F_p \qquad (7.1)$$

Goniometric analysis of the pressure difference (i.e. inside – outside)Δp acting on an element $d\varnothing$ implies that the vertical components of the radial pressure per unit lengthis given by[1]

$$F_p = \int_0^\pi \Delta p \sin \phi R d\phi = \Delta p R \int_0^\pi \sin \phi d\phi = \Delta p R [- \cos \phi]_0^\pi = \Delta p R \ (7.2)$$

Combining equation (7.1) and equation (7.2) then implies that:

$$2\sigma_{circ}t = \Delta pR \qquad (7.3)$$

To calculate the stress in longitudinal direction, the pressure vessel should be considered as illustrated in Figure 7.2. The pressure $d\varnothing$ acting on the circular surface should be in equilibrium with the longitudinal stress acting at the circumference of the vessel.

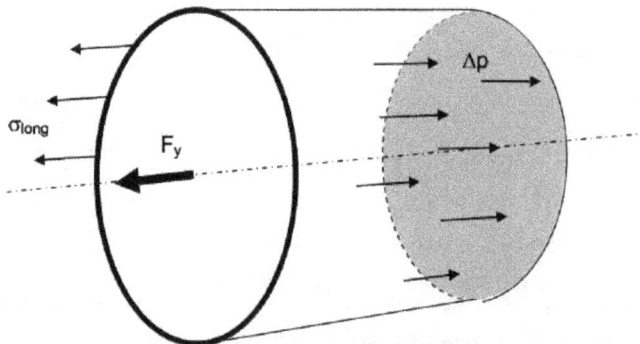

Figure 7.2 Illustration of equilibrium of the internal pressure and the longitudinal stress. (Alderliesten, 2011)

The equilibrium is given by:

$$F_{long} = 2\pi R\sigma_{long}t = \Delta p\pi R^2 = F_p \qquad (7.4)$$

Thus, the circumferential and longitudinal stresses in a pressure vessel can be described with:

$$\sigma_{circ} = \frac{\Delta pR}{t}$$
$$\sigma_{long} = \frac{\Delta pR}{2t} \qquad (7.5)$$

Following from chapter 1 and equation (1.11), the Hooke's law for the bi-axial stress condition (plane stress) in a pressure vessel is then described by:

$$\varepsilon_{circ} = \frac{\sigma_{circ}}{E} - \nu\frac{\sigma_{long}}{E}$$
$$\varepsilon_{long} = \frac{\sigma_{long}}{E} - \nu\frac{\sigma_{circ}}{E} \qquad (7.6)$$

Equation (7.5) implies that the circumferential stress is twice the longitudinal stress. Combining equation (7.5) and equation (7.6) this results in:

$$\varepsilon_{circ} = \frac{\sigma_{circ}}{E}\left(1 - \frac{\nu}{2}\right)$$
$$\varepsilon_{long} = \frac{\sigma_{long}}{E}(1 - 2\nu) \qquad (7.7)$$

For a metallic pressure vessel the Poisson's ratio is about 0.3, which means that with equation (7.7) the circumferential strain is about 4.25 times larger than the longitudinal strain.

7.2.2 Connection between cylinder and spherical end sections

Consider a metallic pressure vessel as indicated in Figure 7.3. In this figure several locations of interest can be identified: the weld lines between the individual metal plates and the transition from a tubular shape to the spherical end sections.

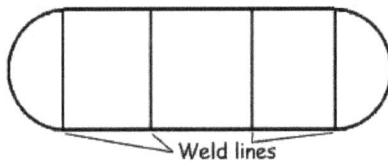

Figure 7.3: Illustration of a pressure vessel containing weld lines. (TU Delft, n.d., 7-3.jpg. Own Work.)

To address the strength of the weld lines in the tubular section, the longitudinal stress should be considered, because that is the stress perpendicular to the weld line pulling the welded components apart. Because the circumferential stress is twice the longitudinal stress, the weld lines should be oriented in circumferential direction, rather than longitudinal direction. Otherwise, the stress loading of the weld would be twice the stress loading of the weld in Figure 7.3 under the same pressurization.

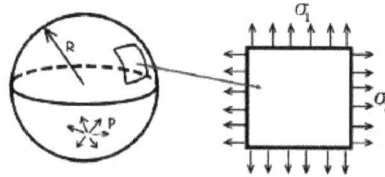

Figure 7.4: Illustration of a spherical pressure vessel. (TU Delft, N.D.)

To calculate the stress in the spherical end sections, first a pressurized sphere is considered, see Figure 7.4. In this vessel the stresses in both directions are equal to :

$$\sigma_1 = \sigma_2 = \frac{\Delta pR}{2t}$$

(7.8)

With the Hooke's law for bi-axial loaded plates, see equation (1.11), this implies for the strains in both directions:

$$\varepsilon_{sphere} = \frac{\Delta pR}{2t}\frac{1}{E}(1-\nu)$$

(7.9)

To create the welded connection between the tubular section and the spherical sections in Figure 7.3, one should consider the question whether the thickness of both sections should be equal or not. In general, discontinuities in circumferential strain as illustrated in Figure 7.5 should be avoided, because they would lead to unwanted deformations, or even damage and/or failure. With the equations for strain given for both sections with respectively equation (7.7) and equation (7.9) , potential strain discontinuities can be assessed.

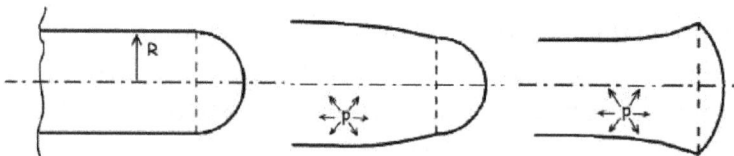

Figure 7.5: Illustration of the transition between tubular and spherical section without strain discontinuity (left) and with strain discontinuities (centre and right). (TU Delft, N.D.)

A similar derivation as presented in section 7.2.1 can be given for this case, except that it will be somewhat more complex. The analysis should be 3D instead of the 2D in Figure 7.1.

To derive the ratio between the thicknesses of both sections the circumferential strain should be equal to the strain in the spherical section, thus $\varepsilon_{circ} = \varepsilon_{sphere}$:

$$\varepsilon_{sphere} = \frac{\Delta p R}{2t} \frac{1}{E}(1 - \nu) = \frac{\Delta p R}{t} \frac{1}{E}(1 - \frac{\nu}{2}) = \varepsilon_{circ}$$

(7.10)

This results in the following relation:

$$\frac{R_{sphere}}{R_{cylinder}} = \frac{2t_{sphere}(1 - \frac{\nu}{2})}{t_{cylinder}(1 - \nu)} = 1 \Rightarrow t_{sphere} = 0.41 t_{cylinder}$$

(7.11)

In case the radii of both sections are identical this means that the thickness of the cylindrical section is 2.5 times the thickness of the spherical section.

7.2.3 Pressurization of an aircraft cabin structure

Of course, the example of a pressure vessel analyzed in the previous section isa simplified case compared to an aircraft fuselage structure under pressurization. Therefore, some remarks have to be made in addition to the previous analysis.

First of all, the thickness of the fuselage shells is not constant throughout the fuselage. Local reinforcement, especially near cut-outs such as doors and windows, will locally redistribute the stresses and change the deformation of the pressured shell structure.

In addition, it was already discussed in chapter 5 and 6 that the fuselage shell is reinforced by stringers and frames to maintain the aerodynamic shape under all operating conditions. Especially the frames have a significant impact on the deformation of the pressurized fuselage and as a consequence on the circumferential stresses. The contribution of the frames to the response of the fuselage to pressurization is illustrated in Figure 7.6.

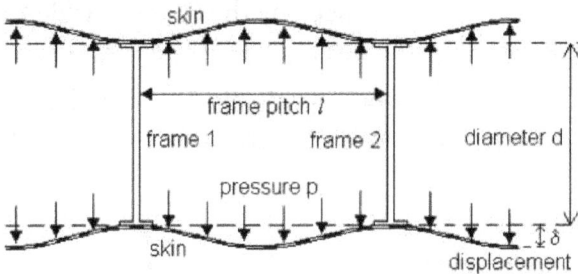

Figure 7.6 Illustration of frame contribution to the fuselage response to pressurization; a pillowing effect occurs due to limited deformation near the frames. (TU Delft, N.D.)

Because failure of the fuselage shell structure under pressurization may imply a disastrous occurrence similar to an exploding balloon, this type of fracture should beavoided at all times. First, the maximum pressurization load is defined to limit the static load that may occur, while secondly, damage retardant features are applied in the fuselage to stop cracks that may occur during over-pressurization.

Limit load cannot be defined for fuselage pressurization according to the formal definition, i.e. once in the lifetime of the aircraft, because the pressurization occurs every flight. For that reason, a larger safety factor is applied to the maximum pressurization load, often a factor 2 is applied.

In case of pressurization loads, the pressure differential is considered to induce the longitudinal and circumferential stress in the fuselage. This pressure differential is the difference between the inside pressure of the aircraft and the outside pressure. The design pressure differential is thus defined as:

$$\Delta p = pd = p_{cabine} - ph \qquad 7.12)$$

where p_{cabin} is the pressure inside the fuselage and ph the pressure at an altitude h. For the cabin pressure often a pressure is taken at an altitude between 2400 and 3000 m, rather than the pressure at sea level. This pressure is considered to be still comfortable for the passengers, while it reduces the design pressure differential for a cruise altitude h.

This can be illustrated as follows. If a safety factor of 2 is applied, the maximum pressure differential is equal to $2pd$. Assume that this maximum pressure differential is equal to 45 kPa. In case the maximum cruise altitude is set to be 9050 m, then a cabin altitude of 2440 m could be used. However, if an aircraft altitude of 10350 m isrequired; this implies a cabin altitude of 3050 m, thus a lower cabin pressure.

7.3 TORSIONAL LOADING OF A FUSELAGE STRUCTURE

Another load case that may act on an aircraft fuselage is torsional loading. As explained in chapter 6, this may occur if the vertical tail of an aircraft is laterally loaded by gust wind, or by forces induced by the rudder. Aside from bending of the vertical tail, this load case will induce rotation of the rear fuselage section.

This section will discuss in more detail the torsional loading of a cylindrical fuselage section and the implications of such load case to the stresses within that structure.

If a torsional moment is applied to a cylindrical shell, stresses occur within that shell structure. To understand the nature of these stresses, one should first consider the shell to be non-continuous as illustrated in Figure 7.7. It is rather evident that the torsional moment will induce the deformation as shown at the left hand side of this figure. To prevent this deformation and to keep the longitudinal edges of the cylindrical shell positioned opposite to each other forces must be applied to these longitudinal edges in the opposite direction as the deformation.

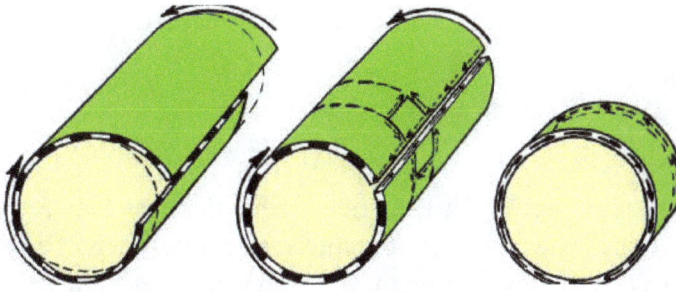

Figure 7.7: Illustration of a cylindrical shell loaded by torsion; in the case the shell is not continuous it deforms as shown left. To prevent that deformation shear stresses must occur. (TU Delft, N.D.)

These forces must be in equilibrium within the shell structures, which implies a shear stress acting in opposite direction as the applied moment, because these shear stresses should create equilibrium with the applied moment.

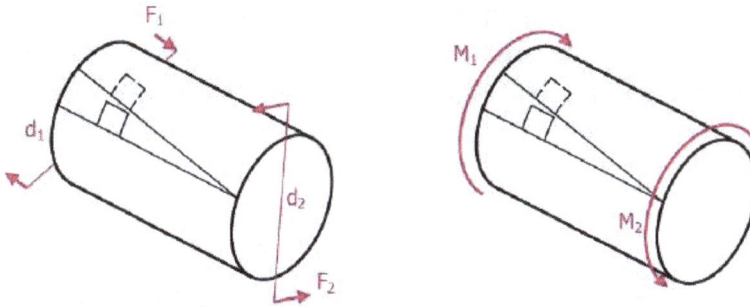

Figure 7.8: Illustration of the force-couple systems and torsional moments acting on a cylindrical shell. (Alderliesten, 2011)

To identify these shear stresses and their direction, one may also look at the deformation of a cylinder deformed under a torsional moment. A rectangular section on such cylinder will deform into a parallelogram, as illustrated in Figure 7.8.

If equilibrium is satisfied in the example shown in Figure 7.8 this means that

$$F_1 d_1 = f_2 d_2 = F_n d_n \Rightarrow M_1 = M_2 = M_n \qquad (7.13)$$

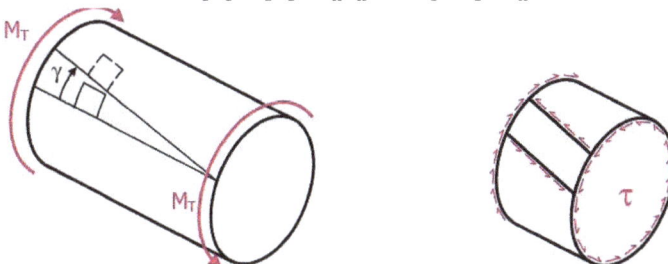

Figure 7.9: Illustration of the resulting shear stresses as a result of the torsional moments acting on a cylindrical shell (Alderliesten, 2011)

Thus a torsional deformation induces shear stress according to equation (1.7), see Figure 7.9

$$\gamma = \frac{\tau}{G}$$

(7.14)

If the shear stress acting in the shell with thickness t at a distance r from the centre of the cylinder over the complete circumference of the cylinder is in equilibrium with the torsional moment M_T (see right hand side of Figure 7.9), this means that

$$M_T = \tau 2\pi\, r\, tr = 2\tau\pi r^2 t$$

(7.15)

Defining the shear flow q as tτ, this equation can be written as

$$M\tau = 2qA$$

(7.16)

In other words, in case a thin walled structure like the fuselage shell structure is loaded by a torsional moment M_T then the shear flow in that shell is solely determined by the enclosed area A. This relationship appears to be independent of the shape of the enclosed area. Thus whether the area is a cylinder as in the example of a fuselage, or a wing box structure, as long as the area is equal, the shear flow q will be the same according to

$$q = \frac{M_T}{2A}$$

(7.17)

This can be illustrated with a wing box as illustrated in Figure 7.10. Dividing the contour of the wing box structure in small elements, then for each element the fractional moment is given by

$$dM_T = q\, ds\, a = q2dA$$

(7.18)

Superimposing all fractions together, one evidently gets equation (7.17).

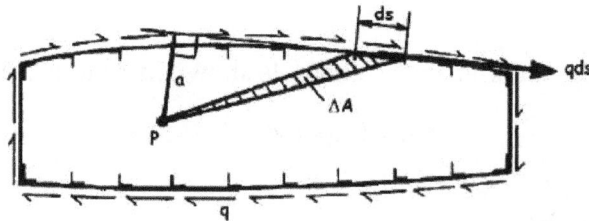

Figure 7.10: Illustration of a torsional moment MT acting on a wing box structure causing shear flow q. (TU Delft, N.D.)

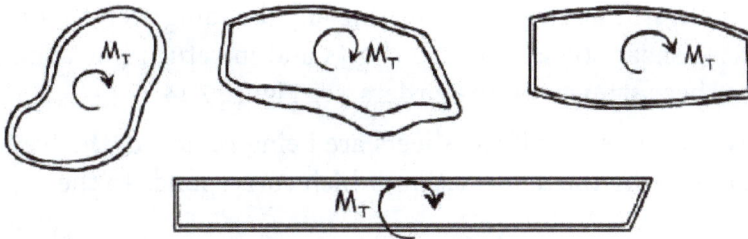

Figure 7.11: Arbitrary enclosed areas for thin walled structures under torsional moment MT. (TU Delft, N.D.)

Thus all structures illustrated in Figure 7.11 have the same shear flow under torsional moment M_T independent of the cross sectional shape. This closed section is therefore often called a torsion box, because it resists torsional deformation efficiently. With the above discussion, it can be concluded that a torsional moment is resisted by shear stresses within the thin walled shell structure. However, in case a cut-out is created within such structure, for example a door or window in an aircraft fuselage, the torsional box will not function well at the location of the cut-out. This is illustrated in Figure 7.12. For this reason cut-outs must be reinforced locally around the cut-out to provide the resistance against shear deformation that was provided by the removed shell structure. This is often created by locally increasing the thickness with doublers and stiff frames and stringers providing rigid corners.

Figure 7.12: Cut-out in cylindrical shell under torsion must be reinforced to provide resistance against shear deformation. (TU Delft, N.D.)

7.4 BENDING OF A WING STRUCTURE

It was explained in chapter 5 that wing bending is taken up primarily by the spars in that wing structure. In this section the function of the spar and its structural elements is explained in more detail and a structural analysis is performed

7.4.1 Shear deformation

First an elementary spar is considered as illustrated in the left hand side of Figure 7.13. The loads acting on that spar structure impose a diamond shaped deformationthat is resisted by the sheets, see section 5.5.2. Assuming that the bars or the girders are stiff and rigid prohibiting any (bending) deformation, then the girders exercise shear stresses on the sheets and in return (for equilibrium) the sheets induce shear stresses on the girders, see Figure 7.14

The consequence is that if the sheets are being removed, the frames will not be capably of resisting the deformation, which corresponds to the explanation in section 5.3.2.

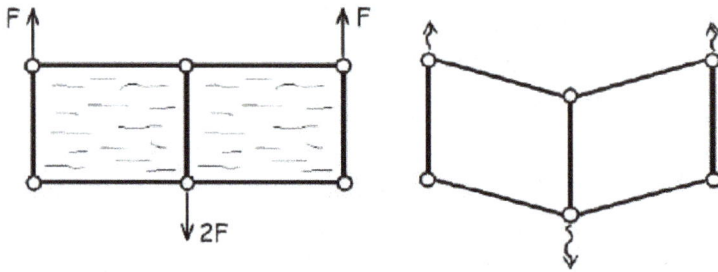

Figure 7.13: Elementary spar constructed from rigid girders and sheets (left) will deform if sheets are removed (right). . (TU Delft, N.D.)

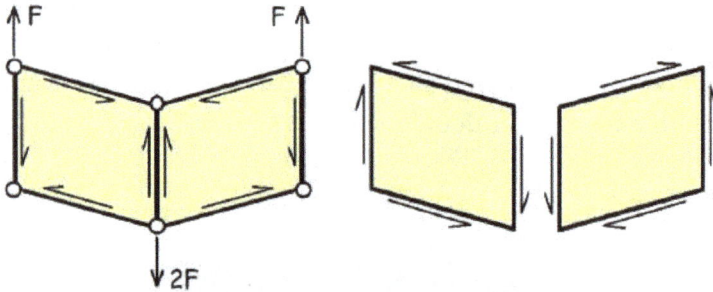

Figure 7.14: Illustration of shear stresses in the girders induced by the sheets (left) and the equilibrium shear stresses in the sheets (right). (TU Delft, N.D.)

The resistance to shear deformation is in fact the resistance to tension and compression under 45 degrees. To relate the shear deformation to the tensional and compressive deformation, the relationship between E,v , G should be used, see section 1.6.

At certain levels of the applied forces F, the shear stress in the sheets will reach a critical level resulting in pleat or wrinkle formation within the sheets. These pleats are often referred to as shear buckling of the sheets. However, one should not consider this as failure, because the diagonal function is still maintained by the sheets. In other words, the structure still functions well, and if the loads are removed the pleats will usually disappear (i.e. when the buckling is elastic).

7.4.2 Stress analysis: breaking-up structural elements

To analyse a spar structure and to calculate the stresses induced by the applied load, the spar must be broken-up into individual elements. An example for a single spar element structure loaded by force F is given in Figure 7.15. For this example, equilibrium must be satisfied, which is used to calculate the reaction forces in the structural elements.

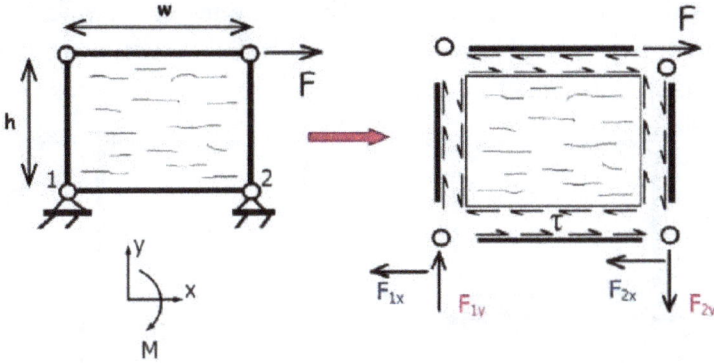

Figure 7.15: Single spar element loaded by force F (left) broken up into individual spars and sheet with reaction forces (right). (TU Delft, N.D.)

For example, equilibrium in horizontal direction yields when:

$$F - F_{1x} - F_{2x} = 0 \qquad (7.19)$$

While equilibrium in vertical direction is satisfied when:

$$F_{1y} - F_{2y} = 0 \qquad (7.20)$$

The relation between the horizontal and vertical forces is given by the moment equilibrium.

Assume that the spar element in Figure 7.15 has a height h and a width w, then equilibrium of moments about location 2 implies that:

$$Fh + F_{1y}W = 0 \qquad (7.21)$$

τSubsequently, the equilibrium in each element can be evaluated. For example, the upper horizontal spar element is loaded at the right hand side by a force F, which can only be kept in equilibrium by the shear forces induced by the sheet over the length w of the spar element. This means that the normal stress at the right hand side of the element is equal to the applied force F and gradually decreases to zero to the left side of the element. Thus the shear stress τ in the sheet is in equilibrium with the normal forces in the frames.

An important remark must be made. Considering loading of elements, then tension and compression imply a physical difference, which easily can be related to a sign convention; tension is positive and compression negative. However, for shear deformation there is physically no difference between both deformations, see Figure 7.16. This means that for the analysis of spar bending a sign convention must be agreed upon like for instance illustrated in Figure 7.16.

Be aware however, that despite tension and compression are easily captured by respectively a positive and a negative sign, this assumes that both cases are identical but opposite of sign. However, physically tension may be considered being different from compression. Consider for example the presence of a crack. Under tension the crack would open and all load has to bypass the crack, while in compression the crack closes, enabling the load to be transferred through the crack.

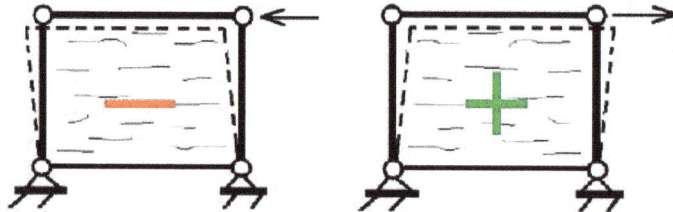

Figure 7.16: A sign convention for shear is needed to analyse the spar deformation. (TU Delft, N.D.)

From the above explanation it is evident that the in case of bending, the shear function of spar webs is essential. However, the webs have to be supported on the upper and lower side by caps or girders in order to realize equilibrium. In this configuration, the webs transfer (external) transverse forces into shear-flows, while the caps transfer shear-flows in normal forces.

7.5 CASE STUDY: BENDING OF WING SPAR

To illustrate the previous explanation, this section discusses the approach for a simplified wing spar loaded in bending, illustrated in Figure 7.17.

7.5 1 Normal and shear forces

The general approach to this problem is to first check the global equilibrium of the problem. This means that the reaction forces at the spar root, i.e. at the clamping area (see Figure 7.18), should provide equilibrium with the external forces F_i.

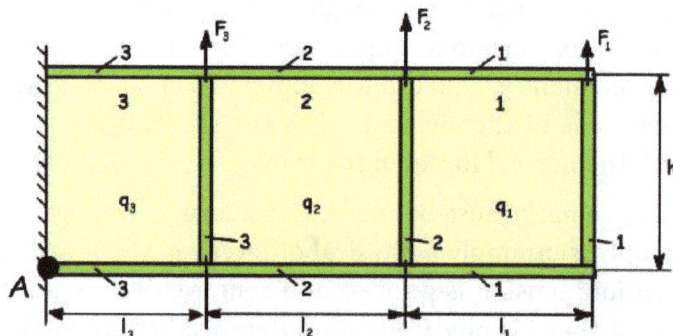

Figure 7.17: Simplified wing spar loaded by forces F1 to F3.
(TU Delft, N.D.)

Global equilibrium implies in (horizontal) x-direction that

$$N_{U4} - N_{L4} = 0 \tag{7.22}$$

And in y-direction

$$F_1 + F_2 + F_3 + q_3 h = 0 \tag{7.23}$$

Equilibrium of moments results in point A

$$N_{U4} h + F1\,(l_1 + l_2 + l_3) - F2\,(l_2 + l_3) - F_3 l_3 = 0 \tag{7.24}$$

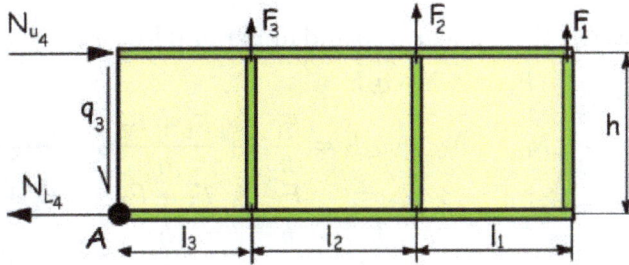

Figure 7.18: Global equilibrium between external forces and reaction forces. (TU Delft, N.D.)

Once global equilibrium is obtained, the problem may be dealt with in detail. The problem is then sectioned into separate shear webs with the related caps. Because the load is applied at the end of the spar, while reaction forces apply to the clamped side of the spar, the analysis will be performed from the spar end in the direction of the spar root.

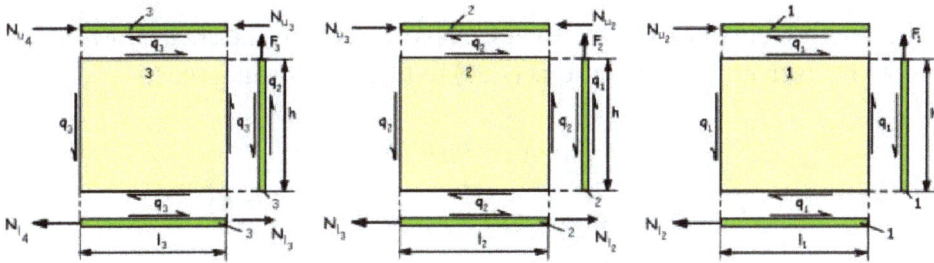

Figure 7.19: The problem of Figure 7.18 broken up into separate sections. (TU Delft, N.D.)

First, vertical equilibrium must be provided between the external load and the shear web. Subsequently, the horizontal equilibrium and equilibrium of moments can be formulated

$$F_1 = q_1 h$$

$$N_{U_2} = q_1 h_1 = \frac{F_1}{h} h_1$$

$$N_{L_2} = q_1 h_1 = \frac{F_1}{h} h_1$$

$$\tag{7.25}$$

Here, q_1 is the shear flow in web plate 1.

Similarly, the equilibrium can be formulated for the second shear web and upper andlower caps:

$$F_2 = q_2 h - q_1 h$$

$$N_{U_3} = N_{U_2} + q_2 l_2 = \frac{F_1}{h} l_1 + \frac{F_1 + F_2}{h} l_2$$

$$N_{L_3} = N_{L_2} + q_2 l_2 = \frac{F_1}{h} l_1 + \frac{F_1 + F_2}{h} l_2$$

$$(7.26)$$

And for the third shear web and upper and lower caps:

$$F_3 = q_3 h - q_2 h$$

$$N_{U_4} = N_{U_3} + q_3 l_3 = \frac{F_1}{h} l_1 + \frac{F_1 + F_2}{h} l_2 + \frac{F_1 + F_2 + F_3}{h} l_3$$

$$N_{L_4} = N_{L_3} + q_3 l_3 = \frac{F_1}{h} l_1 + \frac{F_1 + F_2}{h} l_2 + \frac{F_1 + F_2 + F_3}{h} l_3$$

$$(7.27)$$

From this analysis it can be observed that the transverse shear force is building up towards the spar root. At the first shear web near the wing tip, the transverse shear force D_1 is equal to , while at the second shear web, the transverse shear force $D_2$is equal to F_1 F_2 etc. As a consequence, the transverse shear force at location n (orshear web n) can be generalized as

$$D_n = F_1 + F_2 + \cdots + F_n = \sum_1^n F_n$$

$$(7.28)$$

The relation between the transverse shear force and the shear flow in the web plate can be derived from equations (7.25) to (7.27) according to:

$$F_1 = q_1 h \Rightarrow q_1 = \frac{F_1}{h}$$

$$F_2 = q_2 h - q_1 h = q_2 h - F_1 \Rightarrow q_2 = \frac{F_1 + F_2}{h}$$

$$F_3 = q_3 h - q_2 h = q_3 h - F_2 - F_1 \Rightarrow q_3 = \frac{F_1 + F_2 + F_3}{h}$$

$$(7.29)$$

Thus, the shear flow in each web is described by:

$$q_n = \frac{D_n}{h}$$

$$(7.30)$$

Similarly, it follows that the normal forces in the upper and lower caps at the locationof the vertical stiffener is given by

$$N_{m+1} = \frac{1}{h} \sum_{n=1}^m D_n l_n$$

$$(7.31)$$

As explained in section 6.3.2, the constant shear flow in the shear webs results in an increase in normal forces in the spar caps, because the caps transfer the shear flow into normal forces. Thus equation (7.31) describes the normal forces at the location of the vertical stiffener, the normal forces in-between the stringers require some additional analysis.

Figure 7.20: Upper spar cap of web 2 in Figure 7.19 sectioned at location x. (TU Delft, N.D.)

Consider the upper spar cap in web plate 2 as illustrated in Figure 7.20. Equilibrium for the side of the spar cap indicated by 'A' can be given by

$$N_{U3} - N_{Ux} - q_2 x = 0 \tag{7.32}$$

Similarly, equilibrium for the side indicated with 'B' is obtained with

$$N_{Ux} - N_{U2} - q^2 (l2 - x) = 0 \tag{7.33}$$

With equations (7.30) and (7.31) both equations (7.32) and (7.33) result in

$$N_{Ux} = \frac{D_2}{h}(l_2 - x) + \frac{D_1}{h}h_1 \tag{7.34}$$

This equation illustrates that the normal forces increase linearly from outboard to inboard.

7.5.2 Bending moments

The bending moments at location x in web 2 of the spar can be derived from equation (7.34)

$$N_{Ux}h \equiv M_x = D_2(l_2 - x) + D_1 h_1 \tag{7.35}$$

For web 2 in the spar one can thus obtain that

$$\frac{dM_x}{dx} = -D_2 \tag{7.36}$$

Which is valid on every location x on the spar. Thus equation (7.36) can be generalized as

$$\frac{dM_x}{dx} = -D_x \tag{7.37}$$

7.5.3 Bending of wing spar

Thus once a spar is loaded in bending by external forces F_n the transverse shear forces are known with equation (7.30) and the normal forces in the spar caps are known as

$$N = \frac{M}{h}$$

(7.38)

This enables the determination of the transverse shear force and moment diagram as illustrated in Figure 7.21.

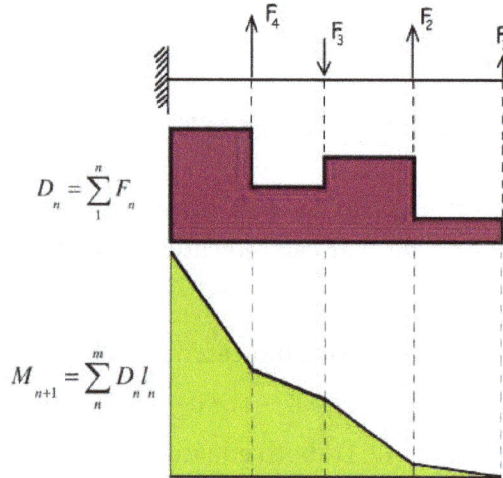

$$D_n = \sum_{1}^{n} F_n$$

$$M_{n+1} = \sum_{n}^{m} D_n l_n$$

Figure 7.21: Bending of wing spar; transverse shear force and moment diagram. (TU Delft, N.D.)

In case of a wing structure, where not only the lift is considered as distributed forces rather than concentrated forces, but in addition also the weight of fuselage and engines is considered, the transverse shear and moment diagrams will look similar to Figure 7.22.

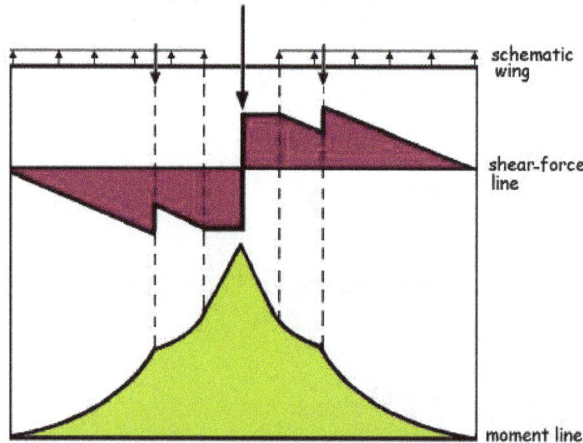

Figure 7.22: Transverse shear force and moment diagram of aircraft wing. (TU Delft, n.d., 7-22.jpg. Own Work.)

It should be noted, that the transverse shear line linearly increases or decreases due to the distributed forces, which implies for the moment line a non-linear increase or decrease.

The shear stress in the webs is given by

$$\tau_x = \frac{q_x}{t_x} = \frac{D_x}{h t_x}$$

(7.39)

While the normal stress in the spar caps is given by

$$\sigma_x = \pm\frac{N_x}{A_x} = \pm\frac{M_x}{h A_x}$$

(7.40)

Here, the parameters hA_x are called the moment of resistance W in a spar.

CHAPTER-8

CONSIDERING STRENGTH AND STIFFNESS

8.1 INTRODUCTION

With the previous chapter in mind, one may understand material behaviour in general, one may know the structural elements of which aeronautical structures are made of, and one may even know about how loads can be transferred through a structure. However, this knowledge may still leave a gap of understanding what aspects are needed to be considered when designing structures and especially materializing them, i.e. selecting appropriate materials for the structure.

To give an impression of aspects to be considered when selecting a material and structural design for a particular application, this chapter will discuss and explain aspects and criteria that relate to material performance in a structural design with a focus on strength and stiffness.

8.2 STRUCTURAL PERFORMANCE

The performance of a structure can be expressed in many different aspects. For example, one may consider the strength at which the structure ultimately fails an important measure for its performance, but if the most dominant requirement is whether the structure under certain loads is limited in deflection, then it may not be the most important performance aspect.

Especially for aeronautical structures weight is considered an important structural aspect. Lightweight design has become an expertise in itself, which seems to aim for structural designs with the lowest possible weight.

Here, one must pay attention to the header above this section, because when discussing weight as structural performance parameter, one should not confuse that with weight or density of structural materials. Although the density of a material will play a certain role in the final structural weight, both categories of weight do not directly relate. Or to put it differently, selecting a material with low density or weight, does not automatically lead to a structure with low weight.

Aside from manufacturing aspects to be considered, the structural performance is a function of

⊙ Properties of materials used in the structural design

⊙ Geometrical features and dimensional aspects of that particular structure

This is illustrated in Figure 8.1. One should be aware, however, that these two categories are not independent of each other.

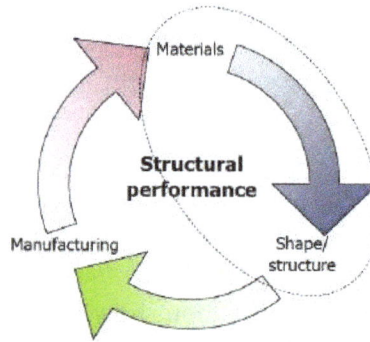

Figure 8.1: Illustration of the optimization considered in this chapter. (Alderliesten, 2011, 8-1.jpg. Own Work.)

For example, an all aluminium aircraft can be designed and manufactured in different ways, where the structural performance is a function of the aluminium alloy properties and the structural geometry. Although balsawood has a significant lower density (and thus weight) than aluminium, designing and constructing an aircraft from that material will lead in the end to a completely different design, in which the geometrical aspects most likely will counteract the benefit of the lower material weight.

Obviously, the strength of balsa wood is significant lower than aluminium which to some extent will play a role in the design, but it is obvious that other aspects like durability and environmental considerations (moisture absorption and subsequent reduction of properties) will also play a role in the selection of materials.

A statement that a certain material has been selected for its low material weight, as reason to obtain low structural weight, testifies therefore for not understanding the concept of structural performance. This will be demonstrated in this chapter with respect to strength and stiffness.

8.3 SELECTING THE APPROPRIATE CRITERION

The optimization of a structural design may lead to different kind of solutions depending on the parameters to which has been optimized. For example, for the same structural application one may obtain different design concepts and even made of different materials, if the concept is optimized for its lowest weight or for lowest manufacturing or operational cost.

But for clarity, the current chapter will primarily focus on the optimization with respect to weight, i.e. the identification of the relevant criteria for the evaluation of mechanical properties of materials. It has been mentioned before (chapter 3) that the strength-to- weight ratio of materials is often considered to select the best material for a structure.

8.3.1 Assessment of material or geometry

First of all, one has to be aware that the lowest weight to be achieved is structural weight and not material weight. This means that if one aims to optimize a structure to the lowest possible weight, not only the material properties and related density have to be assessed, but also the geometrical aspects. This is illustrated in Figure 8.2.

Although this seems to be straightforward, it often leads to confusion. With the introduction of another material in a certain application, the geometry or shape is sometimes also changed. The comparison between old and new design is then often directly related to the introduction of the new material. However, as may be understood from Figure 8.2, one has to distinct between the effect of shape and the effect of the material properties.

However, this distinction is not always easy to make. The example of the steel and aluminium bicycles in section 8.3.3 illustrates that the introduction of a new materialis not always possible with the same geometry; small diameter tubes made of aluminium will not provide enough stiffness to the bicycle frame.

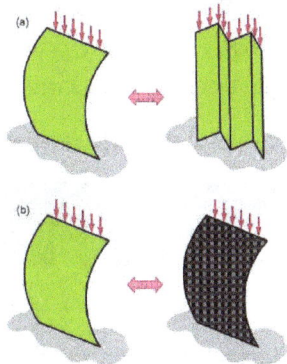

Figure 8.2: Evaluating geometrical aspects for the same material (a) and material aspects for the same geometry (b). (Alderliesten, 2011)

8.3.2 Specific tensile strength

σ_{ult} The specific strength is usually defined as the ratio between strength and density ρ. This ratio relates to another parameter often used in design and construction: the breaking length. The breaking length provides in fact a physical interpretation of the specific strength. Consider a hanging bar, illustrated in Figure 8.3 with a length L, a cross-section A, a failure strength σ_{ult} and a density ρ.

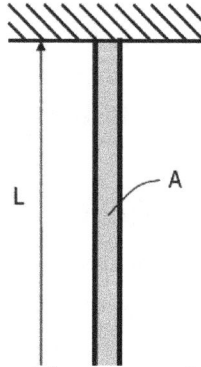

Figure 8.3: Hanging bar with cross-section A. (Alderliesten, 2011)

The weight W in [N] is given by:

$$W = LA\rho g \tag{8.1}$$

where g is the gravitational acceleration. The stress in the upper cross-section is

$$\sigma = \frac{W}{A} = \frac{AL\rho g}{A} = L\rho g \tag{8.2}$$

The length of the bar in Figure 8.3 can be extended as far as the upper cross-sectioncan carry the load. The length at which the upper cross-section reaches its failure strength is defined as the breaking length.

$$L_{ult} = \frac{\sigma_{ult}}{\rho g} \tag{8.3}$$

The unit for the breaking length is usually [km]. A correlation between several materials, their specific strength and the breaking length is given in Table 8.1.

Table 8.1 Correlation between specific strength and breaking length

Material	σ_{ult} [Mpal]	σ_{ult}/ρ [106 Nmm/kg]	ρg [N/dm³]	L_{ult} [km]
Steel AISI 301	1275	159	78.4	16.2
Steel D6AC	1931	248	77.2	25.0
Aluminium 2024 T³	483	174	27.3	17.7

Material	σ_{ult} [Mpal]	σ_{ult}/ρ [106 Nmm/kg]	ρg [N/dm³]	L_{ult} [km]
Aluminium 7475 T761	517	184	27.6	18.7
Magnesium AZ31-H24	290	163	17.5	16.6
Titanium Ti-6AI-4V (5)	950	214	43.5	21.8
Quasi-isotropic CFRP	500	327	15.0	33.3

Example: Specific strength of a simple tension bar

Consider a tension bar to transfer load of 1000 kN from point A to point B over a length of 2 m. Which material (steel or aluminium) provides the lightest solution?

	Steel	Aluminim
Failure Strength	800N/mm²	450N/mm²
Yield Strength	550n/mm²	280N/mm²
Density	7.8kg/dm²	2.8kg/dm³

If permanent (plastic) deformation is not allowed, then the maximum allowed stress equals the yield strength of the materials. The minimum required cross-section is obtained by dividing the load of 1000 kN by the yield strength,

$$A_{steel} = \frac{P}{\sigma_{y,steel}} = \frac{1 \cdot 10^6}{550} = 1818 \text{mm}^2$$

and,

$$A_{alum} = \frac{P}{\sigma_{y,alum}} = \frac{1 \cdot 10^6}{280} = 3571 \text{mm}^2$$

The weight that corresponds to these bars is obtained by multiplying the volume with the density.

$$W_{steel} = \rho_{steel} A_{steel} L = 7.8 \cdot 0.1818 \cdot 20 = 28.4 \text{kg}$$

$$W_{alum} = \rho_{alum} A_{alum} L = 2.8 \cdot 0.3571 \cdot 20 = 20.0 \text{kg}$$

Thus the aluminium solution is 42% lighter than the steel solution.

This could have been evaluated alternatively by directly comparing the performance/weight ratio.

$$\frac{F}{W} = \frac{\sigma_y \cdot A}{\rho \cdot A \cdot L} = \left(\frac{\sigma_y}{\rho}\right) \cdot \left(\frac{1}{L}\right)$$

Because the length L is equal for both cases (thus geometrical aspects are the same, see Figure 8.2), the ratio $\sigma_y \rho$ should be considered. This yields $70.5*10^6$ Nmm/kg for steel and $100*10^6$ Nmm/kg for aluminium. Thus aluminium has the highest specific strength for this case.

The correlation between specific strength and breaking length in Table 8.1 clearly does not provide the reason why aluminium is so often applied in aircraft structures. This is because the specific strength, or more precisely formulated, the specific tensile strength is not the only parameter determining the material to be applied. Depending on the application and the load cases, different parameters have to be considered.

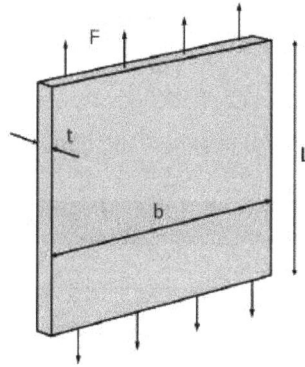

Figure 8.4: Thin sheet loaded in tension. (Alderliesten, 2011, 8-3.jpg. Own Work.) Consider a thin sheet loaded in tension as illustrated in Figure 8.4 made of either AISI 301 steel or magnesium AZ31-H24 (see Table 8.1). The specific tensile properties of both materials are fairly close to each other. The specific modulus of elasticity for these alloys is respectively 2.5 and 2.4 Nmm kg-1.

In particular, the structures applied in aircraft and spacecraft are considered to be thin-walled, i.e. the thickness is relatively small compared to the other dimensions. To illustrate the significance of the specific properties of materials in relation to the application in airframes, two metals are compared.

8.3.3 Specific buckling strength

Thus, concerning the tensile load applied to the sheet, there is no preference for either one of the two metals. However, if compression is considered as load case, sheet buckling has to be considered. The critical buckling strength of a sheet is given by

$$\sigma_{cr} = \frac{Et^2}{Lb}$$

(8.4)

The buckling load is then defined as

$$F_{cr} = \sigma_{cr}bt$$

(8.5)

Thus, the required thickness is calculated with

$$t = \sqrt[3]{\frac{F_{cr}L}{E}}$$

(8.6)

Keeping the buckling load and panel length identical for both metals it is evident that the thickness of the magnesium AZ31-H24 is about 1.6 times the thickness required for AISI301 steel. However, this implies that the weight of the magnesium sheet is about 2.8 times lower than the steel sheet.

In other words, to compare the specific properties for the case of compression buckling, one has to search the highest value for $\sqrt[3]{E}/\rho$.

Table 8.2 Specific buckling strength parameter

Mateial	σult [Mpa]	E [GPa]	$\sqrt[3]{E}/\rho$ [10^6mm$^{7/3}$N$^{2/3}$]
Steel AISI 301	1275	193	0.72
Steel D6AC	1931	210	0.76
Aluminium 2024-T3	483	72	1.50
Aluminium 7475-T761	517	70	1.47
Magnesium AZ31-H24	290	45	2.00
Titanium Ti-6Al-4V (5)	950	114	1.09
Quasi-isotropic CFRP	500	60	2.56

8.4 GEOMETRICAL ASPECTS

The comparison of material properties or, as discussed in the previous section, the specific material properties may indicate the applicability of certain materials for specific structural configurations and load cases. However, one has to be aware that one still has the opportunity to tailor the structure in its geometry, as discussed in section 8.3.1.

To illustrate this aspect, Table 8.3 compares four cross-sections for a beam loaded in bending. The comparison shows that significant weight reductions can be achieved by changing the cross section of the beam, while keeping the material the same. In fact, this is the reason why the aluminium bicycle frames are made of tubes with a larger diameter than the conventional steel frames, see the example in this section.

Table 8.3 Correlation of cross-sectional weight for equal bending stiffness

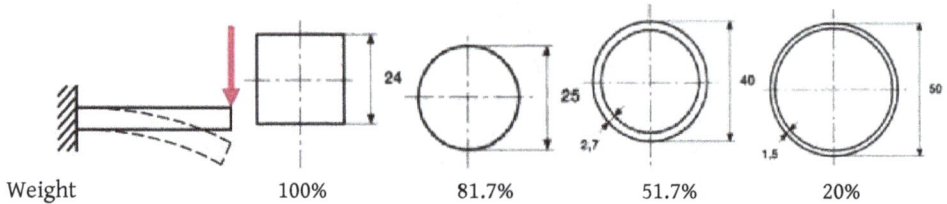

| Weight | 100% | 81.7% | 51.7% | 20% |

In Table 8.4 it is illustrated that when comparing several materials for their specific stiffness that these materials do not always rank highest for all cases. For example, bone material has a lower specific stiffness compared to aluminium, but when the sheet stability is considered, the value for aluminium is lower. This illustrates that the comparison of the materials should account not only for the geometry, but also for the representative loading scenario.

Table 8.4 Illustration of typical specific material properties. (1) Represents weight % of similar loaded structure compared to steel.

Material	Specific modulus		Column stability		Sheet stability	
	E/γ	%(1)	$\sqrt{E}/\gamma$	%(1)	$\sqrt[3]{E}/\gamma$	%(1)
Aluminium	2500	108	9.5	62	1.5	52
Steel	2692	100	5.9	100	0.8	100
Spruce	2340	115	22.3	26	4.7	16
Birch	2538	106	19.8	30	3.9	19
Bone	1500	179	9.0	65	1.6	47
Titanium	2622	103	7.8	76	1.1	73
Isotropic carbon fibre composite	3333	81	14.9	40	2.5	32
Isotropic E-glass fibre composite	536	502	5.1	116	1.1	73
Isotropic aramid fibre composite	1760	153	11.4	52	2.1	38

Example: Steel and aluminium bicycle frames

For many years bicycle frames were made of steel tubes with circular cross section. These tubes were jointed by either lugs or welding. Because of the high strength and stiffness of the steel alloys used, tubes could be used with a relatively small circular cross section. The small circular tubes could be joined easily with lugs, in which the tubes were then brazed to the lug. The alternative joining method is TIG welding which is a straightforward process for steels.

Aluminium, although having better strength-to-weight ratio than steel was not applied, because the applied aluminium alloys could not be welded. With the introduction of weldable aluminium alloys, the introduction of aluminium frames was initially not successful because of fatigue failures. Aluminium usually has a lower fatigue limit than steel, see explanation in chapter 10.

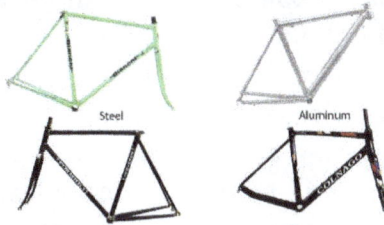

Figure 8.5 Comparison between steel and aluminium bicycle frames. Frame top left: Steel-vintage.com, (2018), CC-BY; frame top right: fietstijden.nl, (2018), Public Domain; bottom left: Saunders, (2008), CC-BY-NC- ND2.0; bottom right: Glory cycles, (2018), CC-BY2.0.

Once reliable and weldable aluminium frames were introduced, many aluminium bikes were sold for their low weight compared to steel bikes. Although the aluminium has a lower stiffness than steel., rigid frames could be achieved by changing the cross-sectional area and shape compared to the steel tubes. Aluminium frames are therefore easily recognized because of the larger tube cross-sections, see Figure 8.5.

8.5 STRUCTURAL ASPECTS

The geometrical aspects discussed in the previous section are important to be considered. The transition in using different materials often comes together with the application of different design concepts. An illustration is provided in Figure 8.6, where the wood, linen and steel trusses of the early aircraft are compared with the stiffenedaluminium shell structure of current commercial aircraft. The concept of load bearing shell structures could only be applied when materials were considered that can be loaded as such.

Figure 8.6 Illustration of the different design related to material usage; wood, linnen and steel trusses (left) and load bearing stiffened aluminium shell structure. Derivative from left: Cliff, (2003), CC-BY2.0 and right: Kolossos, (2006), CC-BY-SA3.0.

Although this seems a very straightforward aspect, often the comparison of materialtechnologies is performed solely by addressing the material properties. For example, the fact that carbon composites in general have lower densities than aerospace aluminium alloys is used to explain that these fibre reinforced materials will result in lighter structures. However, even if similar structures are being considered – for example load bearing shell structures – still the comparison may need attention. Comparing aluminium with carbon fibre composites in a load bearing shell structure, will not automatically lead to similar details design solutions. Where aluminium stringers for example can easily be extruded into preferred shapes, the manufacturing technology for composites may require stiffeners with different geometries. In addition, the difference between aluminium and carbon fibre composites may leadto different selection of shell concept, i.e. shell containing stiffeners and sandwich panel (discussed in chapter 5). The reasons for selecting either one of the two may be completely different for both material technologies.

But even when these aspects are considered, one has to be careful with comparing the materials purely on material properties as determined in a material test. The question here may be what strength of the material should be considered when comparing different technologies. Comparison based on ultimate strength, i.e. , often leads to irrelevant weight estimates, because this parameter is in general not directly decisive for design. Aluminium structure are required not to permanently deform under the maximum occurring loads, which implies that the yield strength often dictates the minimum thicknesses rather than the ultimate strength.

The restriction on strength that can be exploited in composites is even more severe, the maximum allowable strain that may occur in the structure is limited at about 0.35%. This implies a significant reduction in strength to design with.

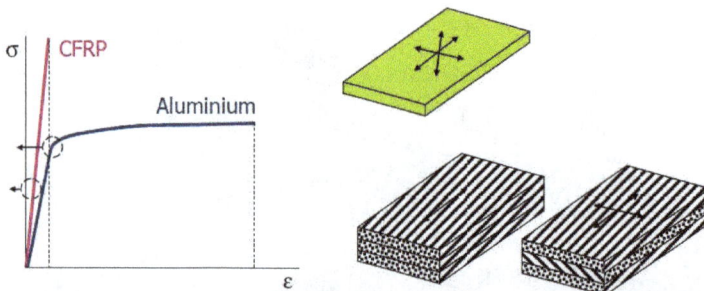

Figure 8.7: Comparison of strength values for different materials (left); high strength linear elastic composites may be limited relatively more than ductile aluminium. The aluminium properties relate to isotropic behaviour, while composite properties often relate to uni-directional composites (right).(Alderliesten, 2011)

Comparing aluminium with carbon fibre composites therefore should be based on equivalent static or fatigue loads. As illustrated here, the static strength comparison may be based on the yield strength in aluminium and about a third of the panel strength (i.e. lay-up of individual plies in the required orientations, not the uni-axial ply strength, see Figure 8.7) in composites.

The actual composite panel lay-up that must be considered depends on the application one has in mind to compare the materials for. As illustrated in Figure 8.8, the composite panel lay-up for a vertical tail will be different from the lay-up needed for fuselage shell panels. But even there, the orientation in the upper fuselage (loaded primarily in tension) may be expected to be different form the side shells (loaded primarily in shear). Some typical values for comparison are provided in Table 8.5.

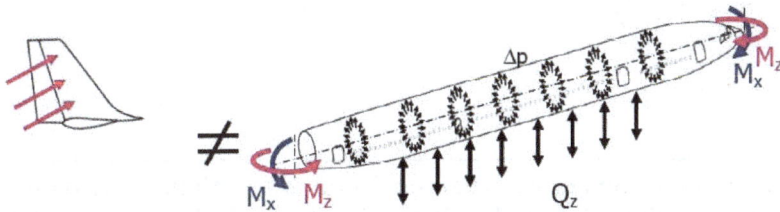

Figure 8.8: The panel lay-up for composite materials is different for the various applications on an aircraft; where empennage structures often allow significant directionality of composites leading to large weight savings, fuselage structures require almost quasi-isotropic laminate lay-ups. (TU Delft, N.D.)

Table 8.5 Comparison between the weight of different cross sections optimized for equal bending stiffness

| Loading mode | Minimum weight for given | |
	Stiffness	Strength
	$\dfrac{\sqrt[3]{E}}{\rho}$	
	$\dfrac{\sqrt[3]{E}}{\rho}$	$\dfrac{\sqrt{q}}{\rho}$

$$\frac{E}{\rho}$$

$$\frac{\sigma}{\rho}$$

$$\frac{E}{(1-\nu)\rho}$$

Table 8.6 Comparison between composites and aluminium based on relevant strength values, with (1) is based on limited to 0.35% design strain and (2) is taken from including notch factor 0.9 as specified in Rice et al., (2003).

Material	Lay-up % 0°/±45°/90°	E [GPa]	E/ρ [MPa m³/ kg]	σ_{max} [MPa]	σ_{max}/ρ [MPa m³/kg]	Comment
CFRP (T800S)	60 / 30 / 10	103	64	360(1)	0.225	Tail plane shells
CFRP (T800S)	40 / 50 / 10	77	48	270(1)	0.170	
CFRP (T800S)	20 / 70 / 10	50	31	175(1)	0.110	Fuselage side shell
2xxx Al-alloy		72	26	440(2)	0.160	
7xxx Al-alloy		72	26	565(2)	0.205	
Al-Li alloy		77	29	515(2)	0.195	

8.6 TYPICAL MISSION REQUIREMENTS FOR SPACESTRUCTURES

The design of the structural components in space structures usually starts with the determination of the mission requirements. However, when it comes to the typical missions for spacecraft and launchers, it is obvious that in general the mass should be minimized as much as possible, while the stiffness and the strength should be ashigh as possible.

In addition, for the launchers yields that they should be able to accommodate the payload and the equipment, and that their mission should be fulfilled with high reliability. The costs of a launching failure are extremely high (estimated at about 180 million Euros for the Ariane 5).

In general, these high demands to the structure imply that structural solutions and materials are often considered that are too expensive and complex for aeronautical structures. Nonetheless, even for spacecraft and launch vehicle structures the requirements are to reduce costs, and to search for the design solution which has proven to have the best manufacturability and accessibility.

8.7 MATERIAL SELECTION CRITERIA

The selection of materials for space applications is an important topic. The stiffness of the structure, and therefore also the materials used, is an important parameter to design against the resonance that may occur during launch of the vehicle.

Oscillations can be either damped or excited, see for example Figure 8.9. These oscillations or vibrations are very important in space structures, because during launch significant vibrations may occur. Examples of vibration problems in aeronautical applications are flutter of main and tail wings.

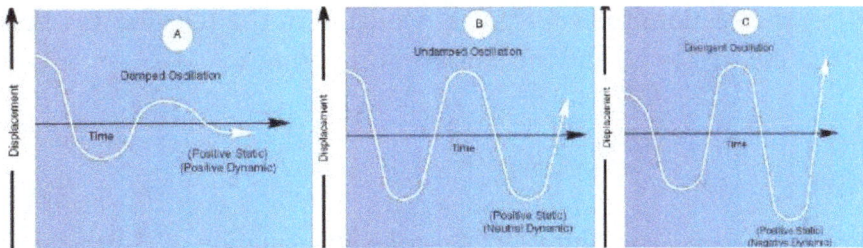

Figure : 8.9: Illustration of the oscillations that are damped (left), not damped (centre) and divergent (right). (TU Delft, N.D.)

Therefore, limiting the natural frequencies of spacecraft is essential to avoid resonance between launch vehicle and spacecraft. In general low dynamic coupling results in lower loads for spacecraft. To dimension the primary structure of a spacecraft (such as a satellite for instance), the first step therefore is to assure that the lowest natural frequency present in the space craft structure is well above the specified minimum frequency by the launcher's user manual (Wertz & Larson, 1999).Once this has been achieved, the structure will be further designed and tailored for all the quasi-static loads that will occur. In this order these steps will be discussed in the following sections.

8.8 STRUCTURAL SIZING FOR NATURAL FREQUENCY

The discussion of structural sizing for natural frequencies, will be explained here using simple examples. Consider the concentrated mass at the end of a clamped beam as illustrated in Figure 8.10. If resonance may occur in both axial and lateral direction, i.e. in x- and y-direction, then loading of the beam may be considered respectively by axial loading of a spring (see introduction of chapter 1) and bending of a beam. For both cases a 'spring constant' k may be determined. For axial direction, the constant k is a function of the stiffness of the spring (represented by EA) and the length of the spring, as is illustrated by Figure 8.12.

$$k_x = \frac{EA}{L} \tag{8.7}$$

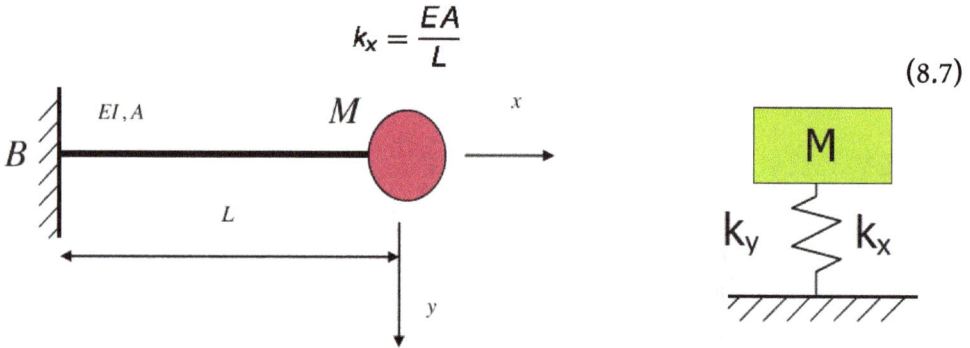

Figure 8.10: Schematic of a concentrated mass at the end of a clamped beam (left) and the simplification of the case to a single degree of freedom case (right). (Alderliesten, 2011)

In lateral direction, the constant k is related to the bending stiffness of the spring (represented by EI) and the length of the spring

$$k_y = \frac{3EI}{L^3} \tag{8.8}$$

The axially loaded configuration, described by equation (8.7), can therefore be related to the case illustrated in Figure 8.11. According to chapter 1, the elongation and strain for this case can be described by

$$\Delta L = P\frac{L}{EA}, \quad \varepsilon = \frac{\Delta L}{L} = \frac{P}{EA} = \frac{\sigma}{E} \tag{8.9}$$

Similarly, the deflection for the laterally loaded case can be given by

$$\delta = \frac{PL^3}{3EI} \tag{8.10}$$

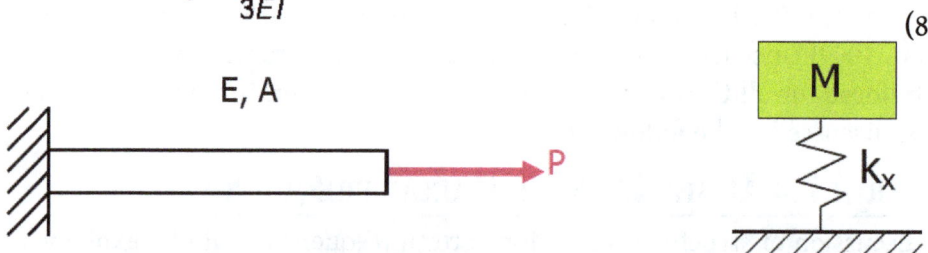

Figure 8.11: Illustration of the beam loaded in axial direction (left) and the simplification of the case to a single degree of freedom case (right). (Alderliesten, 2011)

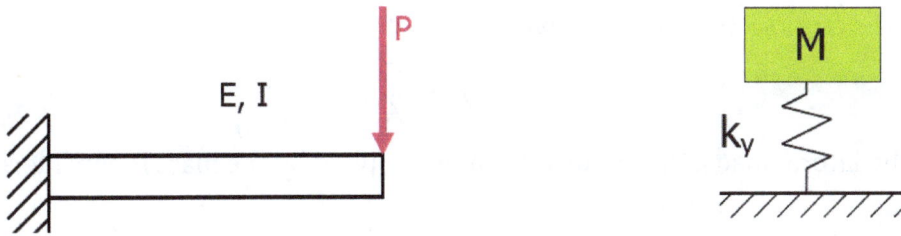

Figure 8.12: Illustration of the beam loaded in lateral direction (left) and the simplification of the case to a single degree of freedom case (right). (Alderliesten, 2011)

The natural frequency in [Hz] is defined as

$$f_n = \frac{1}{2\pi}\sqrt{\frac{k}{m}}$$

$$(8.11)$$

where k is the spring constant and m the mass. If this equation represents the lowest natural frequency that is allowed in the structure, this means for the axial direction that

$$\frac{EA}{L} \geq (2\pi f_n)^2 m$$

$$(8.12)$$

and for the lateral direction that

$$\frac{3EI}{L^3} \geq (2\pi f_n)^2 m$$

$$(8.13)$$

8.9 STRUCTURAL SIZING FOR QUASI-STATIC LOADS

Once the requirements concerning natural frequency are being met, the structure can be designed for quasi-static loads. These loads are directly related to the acceleration of the mass during launch and ascent.

In axial direction, the load is given by the acceleration of the mass in axial (launch) direction with

$$F = mg_x$$

$$(8.14)$$

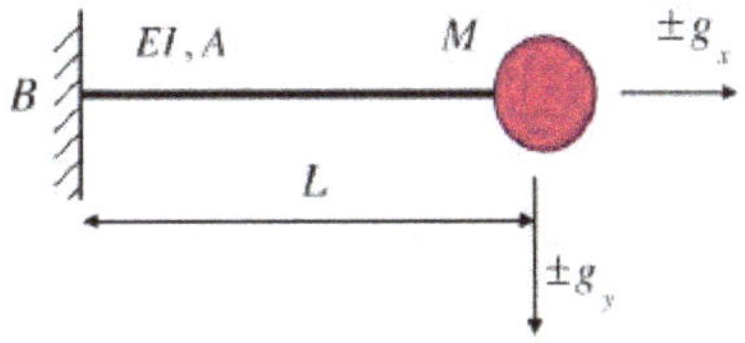

Figure 8.13: Schematic of a concentrated mass at the end of a clamped beam subjected to axial and lateral accelerations. (Alderliesten, 2011)

With the cross section of A this gives in axial direction

$$\sigma_x = \frac{F}{A}$$

(8.15)

For the lateral loads, the bending moment applied by the mass is considered:

$$\sigma_x = \frac{My}{I}$$

(8.16)

Where I is the area moment of inertia for bending, y the distance from the neutral lineto the outer surface of the beam, and M the moment given by:

$$M = FL = mg_x L$$

(8.17)

The allowable stress is the maximum stress that the structure should be capable to sustain without any damage or failure. This maximum stress is calculated by superimposing the stresses due to axial and lateral accelerations, which should be lower than the allowable stress.

$$\sigma_{tot} = \sigma_x + \sigma_y \leq \sigma_{allowable}$$

(8.18)

The allowable stress is the ultimate stress divided by a safety factor.

Another load case that has to be considered is the buckling load applied to the structure by the axial accelerations during ascend. In this case the bending stiffness EI of the beam becomes an important parameter. The Euler buckling load can be calculated with

$$F_{euler} = \frac{\pi^2 EI}{4L^2}$$

(8.19)

Which implies that the loads due to axial accelerations should be limited to:

$$g_x M \leq F_{euler}$$

(8.20)

More on initial sizing of spacecraft can be found in Wertz & Larson (1999).

CHAPTER-9
DESIGN & CERTIFICATION

9.1 INTRODUCTION

With the information presented and discussed in the previous chapters, one may be able to understand the material behaviour in a structure, and one may even be able to calculate based on given load cases the stresses in the various structural components, but that is in itself not sufficient to design, manufacture and operate anaircraft or spacecraft.

Aspects as safety, not easily captured in equations and formulas, must be addressed in certification procedures, to be communicated with airworthiness authorities. Structural design must comply with requirements and specifications, which are often non-negotiable, but sometimes have to be weighted against other criteria or requirements. This chapter describes some of the relevant aspects that relate to material selection, selection of structural design approaches, and certification procedures.

9.2 SAFETY, REGULATIONS AND SPECIFICATIONS

In chapter 5 the functions of a structure have been explained in general when discussing and defining the airframe, and in detail for the various structural elements. To be able to fulfil their functions, structural elements in aircraft and spacecraft must comply with a number of requirements. Several requirements directly relate to the functions to be fulfilled. For example, if an element should be able to carry a certain amount of load, then a requirement may be that any degradation to the element during service should never lead to strength below the minimum strength to carry that load.

However, there are also a number of requirements that do not directly relate to the functions of that particular element, but they may follow from specifications or requirements of the complete structure to which the element belongs. This section will discuss a number of requirements relevant for aircraft and spacecraft structures.

9.2.1 Safety

Safety is considered an extremely important aspect in aviation. Governmental organisations on national and international level have compiled an extensive set of regulations for all aspects concerning aviation. With each accident or even incident reported in the news, the questions on safety are being raised again; is it safe to fly? Can safety be improved?

These discussions are to some extent interesting, because one of the characteristicsof safety is that it is not an aspect easily measured. This is illustrated in Figure 9.1 with a comparison between the number of fatalities of different transportation modes. In general, safety could be associated with the lowest fatality rate. However, to evaluate safety of different transportation modes, or even different flights/missions within only aviation, this fatality rate is considered against a certain parameter. This could be the number of journeys, the number of hours, the distance, or even a combination of these. From the comparison in Figure 9.1 it is evident that safety seems to depend on the way of presentation. In addition to this, there is a psychological aspect contributing to the society's perception on safety. Even if the probability of dying within the air transportation mode is substantially lower than any other mode, and even if the presentation in Figure 9.1 would have resulted in the lowest fatality rate for air transport, then still people may perceive flying as less safe. This is simply related to the fact that with one aviation accident often many fatalities are involved, which receives great attention from the media, while the many car accidents with one or two fatalities per occurrence hardly receive any attention. As a consequence, people are very aware of aviation accidents, but to lesser extent of accidents in othertransportation modes.

aths per billion journeys		Deaths per billion hours		Deaths per billion kilometre	
	4.3	Bus	11.1	Air	0.05
	20	Rail	30	Bus	0.4
	20	Air	30.8	Rail	0.6
	40	Water	50	Van	1.2
	40	Van	60	Water	2.6
r	90	Car	130	Car	3.1
	117	Foot	220	Bicycle	44.6
:le	170	Bicycle	550	Foot	54.2
rcycle	1640	Motorcycle	4840	Motorcycle	108.9

Figure 9.1: Comparison between number of fatalities in different transportation modes relative to respectively journeys, hours and kilometres. (Ford, 2000 and Beck et al., 2007 as listed on Wikipedia – Aviation safety.)

Another drawback of presenting safety by accident rates as illustrated in Figure 9.1 is that this presentation does not distinct between the different routes, operators, distances or type and age of aircraft. Furthermore, the occurrences of accidents in the various transportation modes exhibit large fluctuations, especially in air transportation.

This implies that with the introduction of a new aircraft type into service, a safety level has to be established for that particular aircraft type. But even then, one has to be aware that the operator and the location and routes on which the aircraft is operating may influence the statistics.

The above discussion is based on civil aviation in general and not particularly related to safety of vehicle structures, i.e. structural integrity. Roughly 70% of the accidents is related to human factors rather than structural failures. However, similar considerations may apply. For example, a known factor applied in design of aircraft and spacecraft structures is the so-called safety factor. This factor reduces the load, the number of flights, or another relevant parameter to a level that the probability of failure or accident is reduced to an acceptable level. But what is considered acceptable is not easily measured or defined.

9.2.2 Safety regulations

Structural safety is the joint responsibility of different parties involved. It is not solely the responsibility of the aircraft manufacturer, the operator or the airworthiness authorities. In Figure 9.2 these three parties are presented with their main responsibilities with respect to assuring structural safety. In general, if one of these parties does not comply with its responsibilities, one may assume structural safety to be at risk.

Figure 9.2: Illustration of the three major parties involved in structural safety and their main responsibilities. (Alderliesten, 2011, 9-2.jpg. Own Work.)

9.3 REQUIREMENTS FOR AERONAUTICAL STRUCTURES

Depending on the application or structure and its usage, a variety of requirements can be formulated that have to be met. However, concerning the structure, the requirements relate in general to three aspects:

- ⊙ Strength
- ⊙ Loads
- ⊙ Life time

The strength should not be limited solely to the material strength (i.e. σ_{ult}), but should be considered the resistance to failure. This could be failure of the structure with or without the presence of damage.

Although listed second, the aspect of loads and load cases should be considered themost important and most difficult aspect. The strength of a material or structure canassessed relatively easy. One may either perform tests or perform an analysis basedon the known mechanical properties of the material.

However, to evaluate whether the integrity of the structure can be maintained throughout all potential usage scenarios, one must know the loads related to these scenarios in advance. This implies that the forces acting upon the structure should be known in advance, which is practically impossible.

Often the loads are predicted based upon experience and measurements on earlier aircraft with similar configuration and usage. With the development of analysistechniques based on computational fluid dynamics, loads can be evaluated based on the aerodynamic shape of the aircraft.

Either way, the determination of relevant loads and load cases relates to the events that can be identified during the operational life of the structure. This means that estimation of loads relates to risks, i.e. the probability that certain events may occur and what risks are considered acceptable or not.

The third aspect concerns the topic of structural integrity and durability. The assessment of structural strength and potential load cases could be considered quasi-static. This maximum loads can be estimated and the structural strength to meet these loads can be achieved.

However, most engineering materials applied in aeronautical structures are affected by the environment when exposed sufficiently long, as discussed in chapter 2. In addition, the repetition of loads throughout the operational life may impose additional degradation of the structure, known as fatigue. This topic is discussed in more detailin chapter 10.

Time, however, can be defined in various units. For example, the life of the aircraft can be defined in years, flight hours, or flights. Depending on the component or structure,different units for time may be considered. For example, the pressurization of the fuselage during the flight relates to each flight. The loads related to the pressurization therefore are recorded against the number of flights. However, the usage of the engines is often expressed in flight hours, because the engine components are loaded continuously during the time the engines are operated.

The accurate recording of usage during operational life is important to evaluate whether the aircraft in reality is loaded more or less severe than the loads and load cases considered during the design and certification of the aircraft. Heavier usage ofthe aircraft than anticipated in design may impair the structural integrity before the end of life has been reached.

For the operator the recording against the number of flight hours is relatively easy, as it relates directly to the planning of flights, i.e. each flight in general has a specific duration. However, for certain components the number of flights should be known toassess the actual life.

9.4 STRUCTURAL DESIGN PHILOSOPHIES

Throughout the evolution of flight, structural design philosophies have changed based on experience. Unfortunately, this experience relates to incidents and accidents. Initially, the strength of aeronautical structures was evaluated based on quasi-static loads. The estimation of relevant loads was based on experience and engineering judgements. As a consequence, the load estimation was fairly inaccurate.

Figure 9.3: Illustration of a typical static strength tests on early aircraft. (Fokker, n.d., Public Domain.)

An excellent illustration of the evaluation and experimental assessment of strength of an aircraft structure is given in Figure 9.3. Here a maximum static load is consideredthat is represented in the tests by the number of persons on the wing.

Interestingly in this case the main load on the wing is related to upward bending, while the people in the experiment apply a downward bending of the wing.

9.4.1 Safe life

The strength assessment example of Figure 9.3 implies the assumption that throughout time the structure will remain as it is, i.e. the structural integrity is not affected by either corrosion, accidental damages, fatigue, etc.

The first design philosophy that assumes that the structural integrity is maintained during operational life and that any degradation or strength reduction due to fatigue or corrosion is often denoted as 'safe life'. It can therefore be defined as:

Safe-life of a structure is the number of flights, landings, or flight hours, during which there is a low probability that the strength will degrade below its design strength.

This design philosophy can also be described as safety by retirement. The aircraft or structure is retired at the end of life before structural degradation may impair the structural integrity. This concept is illustrated in Figure 9.4.

As a consequence of the rapid introduction of new aircraft until the 50s of the previous century, this design principle could be considered sufficiently safe. The aircraft were often replaced before the anticipated end of life was reached. However, due to economic reasons aircraft lives were fully used and occasionally extended, increasing the risk of failure during operational life.

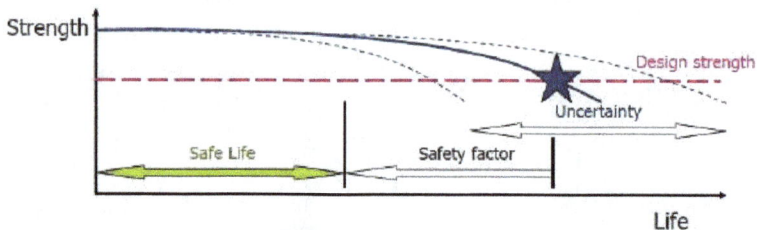

Figure 9.4 Illustration of the Safe life principle; strength reduction is considered beyond end of life. (TU Delft, n.d., 9-4.jpg. Own Work.)

The design philosophy also lead to failure as result of higher loads than anticipated in design and the use of stronger materials with usually poor fatigue properties, crack growth and residual strength. Well known examples are the two Comet aircraft that exploded at cruising altitude in 1954.

9.4.2 Fail safe

Educated by the accidents and incidents, the design philosophy was modified. The structural robustness was increased by adding redundancy to the structure. The design philosophy is referred to as 'fail safe' and can be defined as:

Fail-safe is the attribute of the structure that permits it to retain required residual strength for a period of un-repaired use after failure or partial failure of a principal structural element.

The objective of this design philosophy is that failure of a primary member by fatigueor otherwise does not endanger flight safety. As a consequence, emphasis was put on 'multiple structural member concept'. The redundancy in structural members allowedfailure or partial failure of one member, redistributing the load to other intact structural members, preventing complete failure of the structure. This design philosophy can also be described as safety by design.

The strength evaluation implies therefore that various damage scenarios have to be considered for which the static strength evaluation is performed. In this philosophy, each individual structural item or member is adequately designed according to the safe life concept.

Example: Comet aircraft accidents in 1954

The aircraft accidents occurred after only 1286 and 903 flights. Investigationof the accidents revealed that cracks initiated near the automated direction finder (ADF) window linking up via rivets to adjacent windows, see Figure 9.5. The fuselage was designed with high strength aluminium with poor fatigue characteristics (high notch sensitivity) while relatively high stresses were allowed.

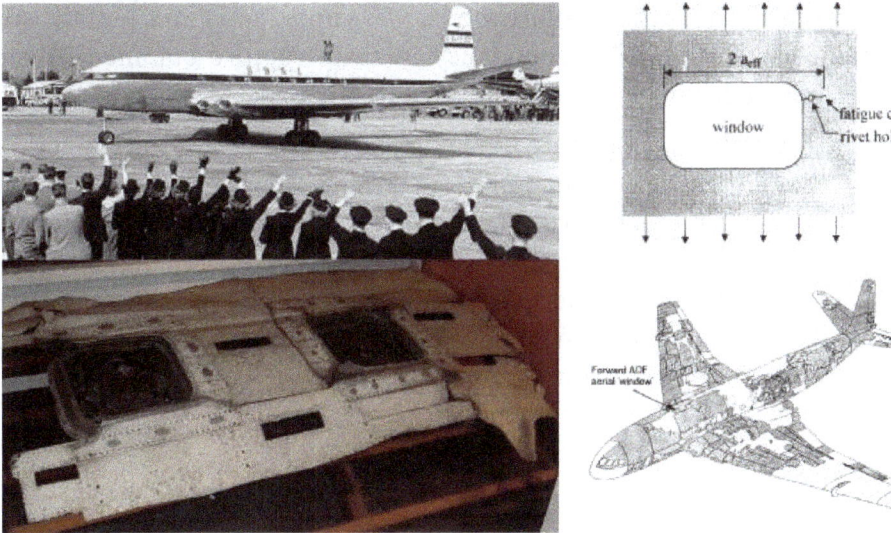

Figure 9.5: Comet aircraft(upper left), illustration of reconstruction of one of the two aircraft (right) and the fuselage section containing the aerial windows. Derivative from top left: British Airways, (1952), Public Domain; top right: Ministry of Transport and Civil Aviation, (1954), Public Domain; bottom left: Krelnik, (2009), CC-BY-SA3.0; bottom right: Ministry of Transport & Civil Aviation (1954), Public Domain.

For certification, a full scale fatigue tests was performed, which only shows the initiation of fatigue cracks after 16000 flights. Subsequent investigation of the fatigue performance of aluminium alloys revealed that the static pressure tests performed before the fatigue load spectrum was applied induced a favourable response of the full scale test article. The static and fatigue test were combined on one aircraft for economic reasons. The effect of the high load induced by two times caused local plasticity near windows and notches, with favourable redistribution of stresses. As a consequence, the full scale fatigue test was un-conservative, i.e. the measure life was longer than the actual life of the aircraft.

Repetition of the full scale fatigue test on a Comet aircraft taken out of service without the static pressurization load revealed initiation of fatigue cracks near escape hatches after 3036 flights. The main advantages of this design philosophy compared to the safe life philosophyare related to safety and economics. The damage could be detected within a given amount of time before full failure occurred, which implies an increase in safety. In the safe life philosophy a structure or component had to be replaced once reaching end of life indifferent of the integrity of the component. In the fail safe philosophy, a structural member could be kept in service until partial failure occurred or damage was observed.

Although the fail safe design philosophy implied an increase in safety, still incidents and accidents occurred induced by structural failures. Evaluating these failures revealed that not all failure modes were anticipated in the static strength evaluation. In addition, the redundancy in the structure obtained by multiple elements did not consider particle failure of multiple elements.

For example the lug illustrated in Figure 9.6 has been considered for decades as the typical illustration of the fail safe concept. However, once one of the lug elements contains a crack, its stiffness reduces, redistributing the load to the other lug members. As a consequence, all members of the lug start cracking simultaneously. The occurrence of multiple cracks in adjacent components or elements is called Multiple Site Damage (MSD). In case of MSD, the fail safe design philosophy becomes ineffective.

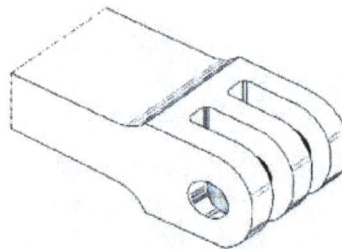

Figure 9.6: Typical illustration of fail safe redundancy in a lug. (TU Delft, n.d. 9-6.jpg. Own Work.)

9.4.3 Damage Tolerance and durability

Since 1978 the aviation requirements (FAR/JAR) adopt the damage tolerance philosophy. This philosophy can be defined as:

The ability of the structure to sustain anticipated loads in the presence of fatigue, corrosion or accidental damage until such damage is detected through inspections or malfunctions and is repaired. The damage tolerance design philosophy is not considered a replacement of the safe life and fail safe philosophy, but rather an advanced concept that combines these two into a new philosophy.

The main advantages of the philosophy are twofold. First, it is assumed that defects, flaws and imperfections are present in the structure directly after manufacturing. These flaws and defects may increase during operational life inducing degradation of the load bearing capability of the structure. Second, the damage (fatigue corrosion, impact) may be present in the structure and even grow until detected during prescribed inspections and subsequently repairs. This repair assumes that the structure is restored to its original strength.

The damage tolerance design philosophy can also be described as safety by inspection. The determination and execution of regular inspections forms an inherent part of the aircraft design.

Example: Aloha airlines accident in 1988

In 1988 a Boeing 737 operated by Aloha airlines lost a large portion of the upper fuselage during flight. Fortunately, all passengers were tied to the chairs with their belts limiting fatalities to one flight attendant.

The investigation of the accident revealed that the riveted lap joints were susceptible to corrosion and contained multiple cracks (MSD). The operators were informed by the aircraft manufacturer about the susceptibility to fatigue and corrosion, especially for warm, humid and maritime air environment near Hawaii, but the operator did not perform sufficient inspections.

This example could be taken as an example that even the damage tolerance philosophy does not guarantee flight safety. Although part of design, inspection and repair have to be performed in order to limit structural failures.

Figure 9.7: Photos of the damage to the fuselage of the Aloha airlines aircraft. Derivative of NTSB, (1989), Public Domain.

The damage tolerance design philosophy as currently applied to aeronautical structures is closely tied to the durability concept. Durability can be defined as:

The ability of the structure to sustain degradation from sources as fatigue, corrosion, accidental damage and environmental deterioration to the extent that they can be controlled by economically acceptable maintenance and inspection programs. The combination of the damage tolerance concept and the effect of environment on structural integrity implies that the damage scenarios considered in the strength evaluation should account for the superposition of cases, i.e. fatigue in metals together with corrosion, fatigue delamination in composites together with reduced resistance due to moisture absorption.

9.5 DESIGN APPROACH

The approach applied in designing aeronautical structures is to identify all critical structural locations for which detailed evaluation must be provided. For each of these locations it must be determined whether inspection is possible or not. This approach is illustrated in Figure 9.8. Currently, only the landing gears and attachments are certified according to the safe life design philosophy, because these components are considered practically impossible to inspect with inspection intervals sufficiently long to comply with durability requirements.

The strength justification of the structural elements is to great extent based on experimental substantiation. In fact, the requirements specify that it must be shown with sufficient tests (supported by analysis) that the probability of failure is negligible. Figure 9.9 illustrates that the aircraft manufacturer performs many generic tests that are not specifically for an aircraft, but applicable to any of the aircraft it considers developing. The data from these tests form the basis for the strength evaluation. Throughout the development of the aircraft, more detailed and complex tests are performed to evaluate and justify the behaviour of the actual structures. Near the end of the development, component tests and full scale fatigue tests can provide the basis for certification.

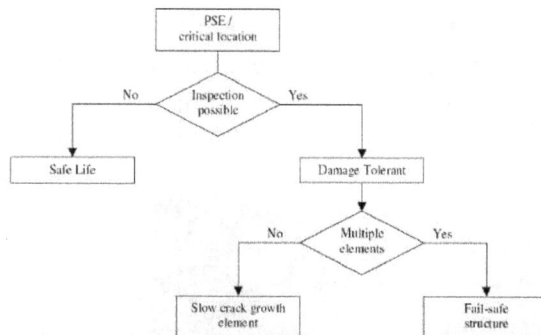

Figure 9.8: Schematic presentation of the design approach. (Alderliesten, 2011, 9-8.jpg. Own Work.)

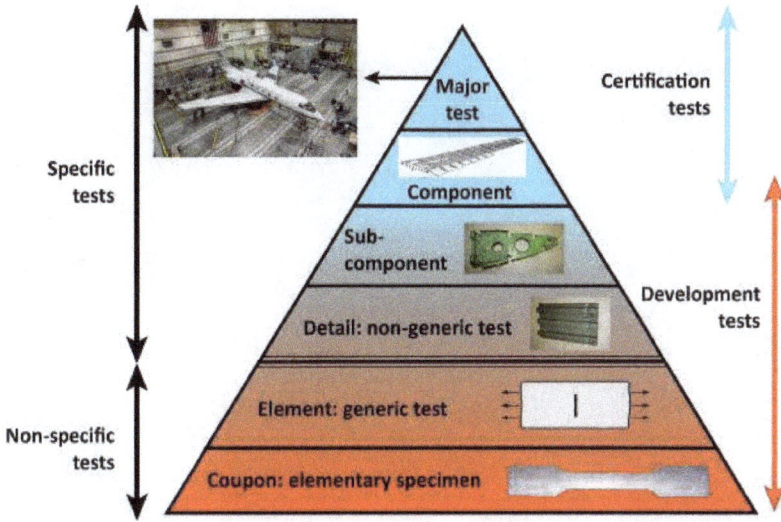

Figure 9.9: Illustration of the experimental pyramid.Derivative from top row: Ulbrich-NASA, (2014), Public Domain; second row: DTom, (2007), Public Domain, third row: Saunders-Smits, (2018), 9-9-c.jpeg, Own Work; fourth row: Saunders-Smits, (2018), 9-9-d.jpeg, Own Work; fifth row: Yapparina, (2014), CC-BY-SA3.0; Bottom row: Wizard191, (2010),

CHAPTE-10
FATIGUE & DURABILITY

10.1 INTRODUCTION

The topic of fatigue and durability is an important topic especially for aeronautical structures that are designed to be operated for several decades while assuring the structural integrity. Based upon the discussion in chapter 2, one can imagine that thematerial properties will change over time due to ageing of the material. Of course this can be accounted for by setting the appropriate allowables based upon the virgin material strength reduced by knock-down factors.

However, although that represents a considerable part of the durability aspect, it will be insufficient for assuring structural integrity. The development of different design philosophies, presented in chapter 9, has proven that structural safety can only be assured with a coherent approach for the design and manufacturing phase, and subsequent operational life.

An important aspect in this sense is the occurrence of fatigue and fatigue damage in structures that are repetitively loaded in service under various loading and environmental conditions. It will be explained in this chapter that fatigue cannot be treated independent of environmental aspects, part of the durability assessment. Therefore, fatigue and durability are discussed in relation to each other in this chapter.

There are some remarks to be made here before stepping into the details. First, fatigue as a damage phenomenon has been discovered primarily in metallic structures (steel in train applications, aluminium in aircraft applications). What has been understood is that from the decades of experience and research is that repetitive (tensile) loading may lead to initiation and propagation of cracks that at some point in time may lead to component failure. It has also been understood how this aspect should be treated in design, which has led to the development of methods fordesigning against fatigue.

What seems to be not well understood is that the relation between repetitive or cyclic loading and initiation or formation of damage applies to all engineering materials and material categories, but that the nature of damage and thus the relevant cyclic load cases may be different from one material category to another. This will be further explained in section 10.4.6.

Another aspect requires more awareness is that designing against fatigue is not an additional task one should perform in the detail design phase. Designing against fatigue follows from the correct mind set of an engineer and should be incorporated already in the early conceptual and preliminary design. If fatigue has not been considered in the early design phases, the design work in the detail design phase willbe merely repairing a fatigue sensitive structure.

10.2 STRESS AND STRAIN CONCENTRATIONS

10.2.1Definition

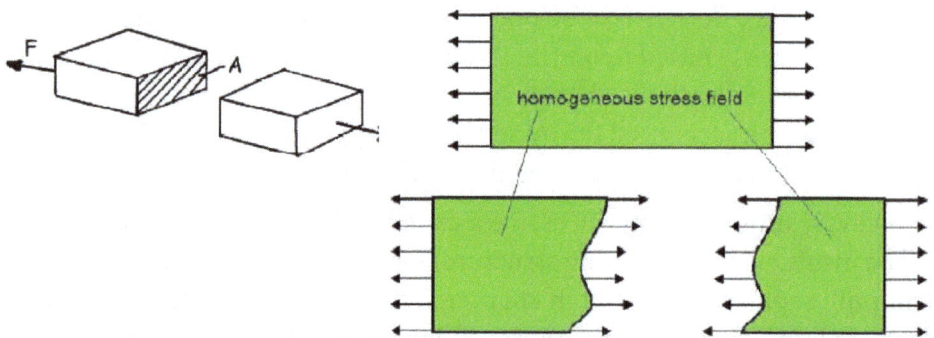

Figure 10.1: Illustration of evenly distributed stresses in a homogeneous material. (TU Delft, N.D.)

Following from the definitions provided in chapter 1, the stress in a material or component is basically calculated by dividing the applied force by the area. Or in equation form:

$$\sigma = \frac{F}{A}$$

$$(10.1)$$

This is considered correct if the material is homogeneous and in undisturbed in shape or geometry. This concept is illustrated in Figure 10.1. Independent on where a cross- section is considered, the stresses will be equal and evenly distributed through the material. These stresses can be calculated with equation (10.1) .

However, if the geometry or the shape is changed, or if a cut, damage or crack is created in the material, then this forms a disturbance for the stress field. The load that has to be transferred through the material, illustrated in Figure 10.2 with stress lines, has to bypass the cut or damage.

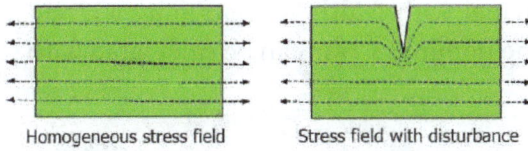

Figure 10.2: Comparison between evenly distributed stress field (left) and a disturbed stress field (right). (TU Delft, N.D.)

Assuming that equation (10.1) is still valid, one should at the location of the disturbance calculate the stress by dividing the applied load F by the reduced cross section. Because the cross-sectional area is reduced while the load remains the same, this implies that the stress at the location of the disturbance is higher than in the undisturbed stress field. This higher stress is generally denoted as the nominal stress and is defined by

$$\sigma_{nom} = \frac{F}{A^*} = \frac{A}{A^*}\frac{F}{A} = \frac{A}{A^*}\sigma_{gross} \qquad (10.2)$$

This would suggest that if the failure of a disturbed or cracked component has to be predicted, one may assume that once the nominal stress σ_{nom} equals ultimate strength of the material, failure would occur. However, before this equilibrium has been reached, the material or component will fail.

This phenomenon is related to the fact that a concentration of stress is present at the tip of the notch. This is illustrated with the higher density of stress lines at the notch tip in Figure 10.2. The undisturbed and evenly distributed stress field has to bypassthe notch, which at the location of the end of the notch creates a high concentration of stress. This phenomenon is indicated with the term stress concentration. To evaluate the stress concentration, a stress concentration factor K_t has beendefined

$$K_t = \frac{\sigma_{peak}}{\sigma_{nom}} \qquad (10.3)$$

Here the peak stress is the highest stress right at the edge or tip of the notch and thenominal stress$\sigma_{nom}$is calculated with equation (10.2).

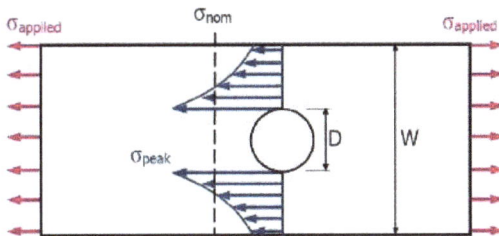

Figure 10.3: Example of a flat sheet containing a hole loaded by a far field stress. (Alderliesten, 2011)

Consider a flat panel with width W containing a circular hole with diameter D, see Figure 10.3, the nominal stress can be calculated with

$$\sigma = \frac{W}{W-D}\sigma_{applied}$$

(10.4)

The stress concentration factor can then be calculated with equation (10.3). The only question at this point will be whether the peak stress is known. Because people in the past did not have the availability of dedicated software that can be used to calculate stresses at specific locations for any geometry under load, like with for example Finite Element Software, theoretical solutions were mathematically derived.These theoretical solutions were derived for the case of notches in sheets with infinite dimensions, because mathematically the boundary constraints for this case are exactly known.

For the finite geometries, subsequently correction factors were determined based upon calculations or experimental data. These solutions were combined in handbooks, like for example Peterson's Stress concentration factor (Pilkey, W.D., 1997).

A well-known general solution for an elliptical hold in an infinite sheet illustrated in Figure 10.4 is given by:

$$K_t = \frac{\sigma_{peak}}{\sigma_{nom}} = 1+2\frac{a}{b} = 1+2\sqrt{\frac{a}{r}}$$

(10.5)

This means that the stress concentration factor for a circular hole in an infinite sheetis equal to K_t = 3. This value is indifferent from the hole diameter. The relation between the peak stress at the notch and the nominal stress in the remaining cross-section section remains 3, independent what hole diameter is considered.

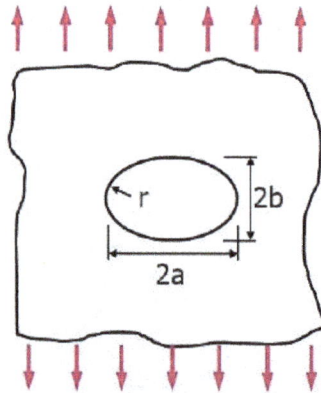

Figure 10.4: Illustration of an elliptical hole in an axially loaded infinite sheet. (Alderliesten, 2011)

An important remark has to be made. When introducing the stress concentration factor K_t and the peak stress at the notch edge, it was argued that the component would fail before the nominal stress σ_{nom} would have reached the ultimate strength of the material.

Implicitly, this explanation and subsequent derivation may lead to the assumption that failure would thus occur when the peak stress σ_{Peak} reaches the ultimate strength. This is correct for materials that remain (approximately) linear elastic until failure, like for example fibre reinforced polymer composites.

However, metallic materials usually plastically deform beyond the yield strength. In that case the behaviour will be different. This is illustrated in Figure 10.5. The stiffness of material reduces significantly beyond yielding (evident from the tangent slope compared to the linear elastic slope, represented by the Young's modulus E). As a result of this local stiffness reduction, stress will be redistributed to the remaining material.

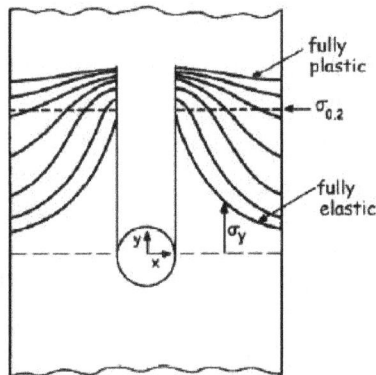

Figure 10.5: Effect of elastic-plastic material behaviour on the stress concentration. (TU Delft, N.D.)

The example given on the next page, illustrates that linear elastic materials have significant higher notch sensitivities than ductile materials. As a result, the failure strength of a notched component made of linear elastic materials is usually lower than that of equivalent elastic-plastic materials.

On the other hand, as will be discussed later in this chapter, composites are usually less sensitive to fatigue loads. In other words, comparing aerospace aluminium alloys with carbon fibre composites, one may conclude that stress concentrations are detrimental for static loads on composites and detrimental for fatigue loads in aluminium alloys.

10.2.2 Saint Venant's principle

An important principle for structural analysis is the principle of Saint Venant. This principle states that the effect of a local disturbance, due to for example a notch, remains limited to the direct neighbourhood of the location of that notch, see Figure 10.6.

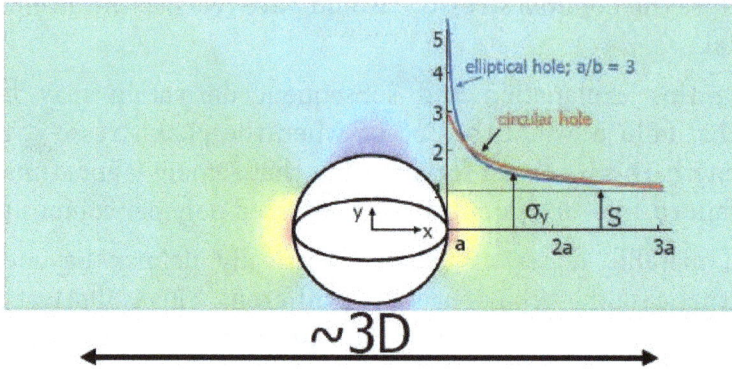

Figure 10.6: Illustration of Saint Venant's principle: Stress disturbance due to a notch remains limited to the direct neighbourhood of the notch (Alderliesten, 2011)

Example: Effect of plasticity on

Consider a straight sample containing a hole, as illustrated in Figure 10.5. If the material behaviour would be linear elastic until failure, like for example carbon fibre composite, this sample would fail when , which means that the gross strength of failure is equal to

$$\sigma_{applied} = \frac{W}{W-D}\sigma_{nom} = \frac{\sigma_{ult}}{K_t}$$

For a sheet with infinite dimensions, this means that the failure strength is equal to. For isotropic carbon fibre composite the failure strength can be about 500 MPa, which means that failure of the sample would occur at 166 MPa.

If the material behaviour would be elastic-plastic like for example aluminium

$$\sigma_{applied} = \frac{W}{W-D}\sigma_{nom} = 0.9\sigma_{ult}$$

For a sheet with infinite dimensions, this means that the failure strength is equal to about 0.9 . With the failure strength of aluminium 2024-T3 of about 450 MPa, this implies failure would occur at about 405 MPa.

The benefit of this principle is that the stress analysis of a loaded structure can be divided into a

- Global stress analysis (considering the structure as a whole)
- Local stress analysis of all details (containing notches)

This approach is justified by the Saint Venant's principle; a structure faces the same stresses due to applied loading indifferent of the presence of a notch at a certain location, except for the location of the notch itself.

10.2.3 Stress distribution around a hole

For the sake of simplicity, the stress concentration factor K_t has been introduced while limiting the discussion to the stress in the net section (nominal stress) and the stress at the notch edge (peak stress). However, it is obvious that the disturbance in stress created by the presence of the notch is somewhat more complex. The peak stress described by K_t is only at the location in the net section. Away from the hole edge or along the contour of the notch, the stresses reduce and may even become negative.

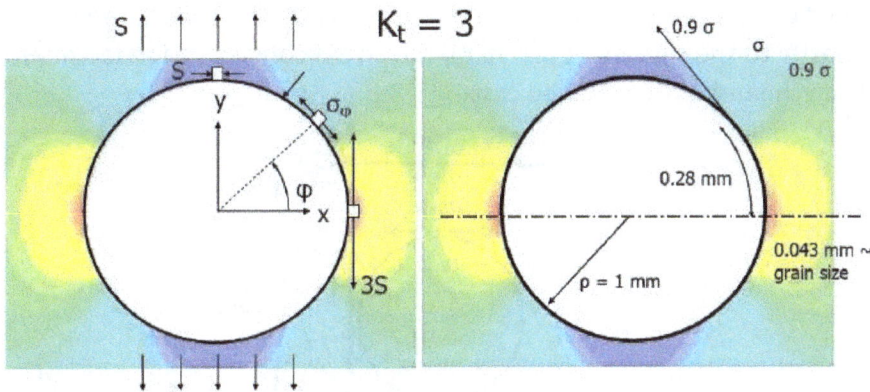

Figure 10.7: Illustrations of the tangential stress distribution around a hole in an infinite sheet; tensile peak stress at $\phi= 0, \pi$ changes to compressive stress at $\phi= \pm \frac{1}{2}\pi$ (left); drop in stress away from the hole is more severe than along the notch contour. (Alderliesten, 2011)

This is illustrated in Figure 10.7. In this figure a peak stress of 3S is present at both sides of the hole, while a compressive stress is present at the locations given by $\phi= \pm \pi/2$. This compressive stress may be understood when considering the stress lines similar to the ones illustrated in Figure 10.2. These stresses may be decomposed in stresses in horizontal and vertical direction, where the horizontal component just above and below the hole induce a compressive stress.

Another aspect that can be observed in Figure 10.7 is that the stress drops quickly moving away from the hole, while the drop along the contour is significantly less. This implies there is a certain area along the notch contour that may be considered relatively high stressed (in the order of 0.9), where potentially fatigue cracks may initiate.

10.2.4 Superposition principle

According to the definition given in section 10.2.1, the stress concentration factor K_t is linearly dependent on the stress. Because stresses may be superimposed, the stress concentration factor K_t coming from a complex set of stresses may be derived by addressing the different stress systems separately and subsequently superimposing the (peak) stresses at the notch.

This is illustrated for a bi-axially loaded infinite sheet containing a hole in Figure 10.8. To calculate the stress concentration factor K_t both stress systems σ_1 and σ_2 may beconsidered separately. For the stresses in direction σ_1 a peak stress of $3\sigma_1$ may be observed at the hole edge as indicated in the figure. However, looking at the stress then a compressive stress equal to $-\sigma_1$ is present at that location (see Figure 10.7). The actual peak stress at the considered location is therefore less than $3\sigma_1$.

In other words, the stress concentration factor for uni-axially loaded sheet containing a hole is higher than the same sheet bi-axially loaded *in tension*. If the sheet would be bi-axially loaded by a tensile stress σ_1 and a compressive stress σ_2 (thus opposite to the sign in Figure 10.8), the stress concentration factor would be higher than the uni-axially loaded case.

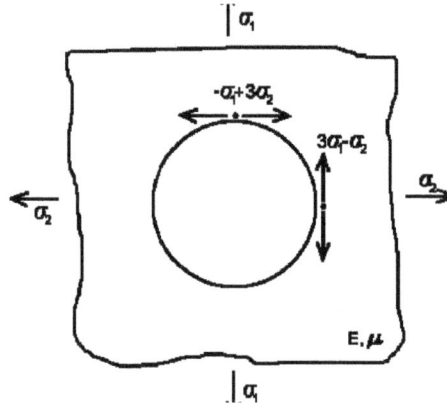

Figure 10.8: Illustrations of bi-axial loading of a circular notch in an infinite sheet. (TU Delft, N.D.)

10.3 REINFORCEMENT OR WEAKENING?

10.3.1 Train axles

The significance of the theory on stress concentrations can be illustrated with the example of train axles. Train axles are heavily loaded components that due to the high rotational speed are repetitively loaded in (reversed) bending. One could consider reducing the load by reinforcing the axle. Reinforcement implies that the axle becomes stronger, but after quite an amount of train axle failure in the 19th century people became aware that reinforcing may imply a weakening.

Example: Versailles railway accident

On the 5 October 1842 a disastrous accident occurred near Versailles. This isthe first major train accident reported in history. The accident occurred due toa fatigue failure of a locomotive front axle, causing collision and derailing of the locomotives. In the subsequent fire about 60 people died.

The investigation into the cause of this accident was performed by Rankine. He recognized distinctive characteristics of fatigue failure and the danger of stress concentrations that locally raise the stress.

The Versailles accident and the subsequent railroad accidents led to intensive investigations of train axle, coupling and rails failures due to fatigue. Almost within a decade after 1842 over hundreds of broken axles had been reported.

It is important to realise the significance of this accident and the magnitude of fatalities. About the same number of fatalities occurred in the two Comet accidents in 1954 a century later, see chapter 9. These two accidents still are considered a major milestone in the history of aviation and the initiator of the investigations into the fatigue phenomena.

Figure 10.9: Illustrations of the railway accident, the train axle and investigator W.J.M. Rankine. Derivative from top left: Unknown Artist, (1842), Public Domain; middle: Unknown Artist, (1842), Public Domain; top right: Glyn, (1844), Public Domain; bottom right: Anon. (N.D.), Public Domain.

This phenomenon can be illustrated with Figure 10.10. The evenly distributed stress in the section of the axle with diameter d has to change into a lower evenly distributed stress in the section of the axle with diameter D. This disturbance of the homogeneous stress field in the radius of the thickness step causes a concentration of stress with a peak stress at the radius contour.

The radius becomes therefore important; a small radius induces a high stress concentration, while a large radius reduces the stress concentration significantly. Often this implies some engineering, because very large radii are not always possible.

Table 10.1 Effect of the radius on Kt for the axle in Figure 10.10 with D/d=2

r/D	0.04	0.1	0.25
K_t	3	2	1.5

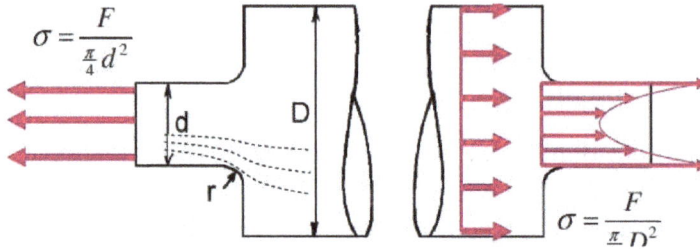

$$\sigma = \frac{F}{\frac{\pi}{4}d^2}$$

$$\sigma = \frac{F}{\frac{\pi}{4}D^2}$$

Figure 10.10: The effect of increasing the thickness of an axially loaded axle. (Alderliesten, 2011)

10.3.2 Repairs

The example discussed in the previous section should be seriously considered, because it relates to a general problem. If a structure in general is reinforced with additional material or structure, the influence of the geometry may induce unwanted concentration of stress. This may also be the case when for example an aircraft structure is reinforced with doublers or when such structure is repaired with patch repairs. The thickness step related to the doubler or patch implies a location for stress concentrations. This may be even more important when considering the higher stiffness of the reinforced area.

This is illustrated with examples in Figure 10.11. Especially the illustration in this figure highlights using similar stress lines as in Figure 10.2, that the stiffer area of the patch repair attracts more load than originally would be present at that location in the structure. The patch may successfully reinforce the damage it originally repaired, butthe higher stress with the thickness step induced stress concentration may shift the critical areas to the edge of the patch repair.

A solution to prevent a failure at this location may be to taper the edge, as illustrated with the bonded Glare repairs on the C-5 Galaxy (right hand side of Figure 10.11).

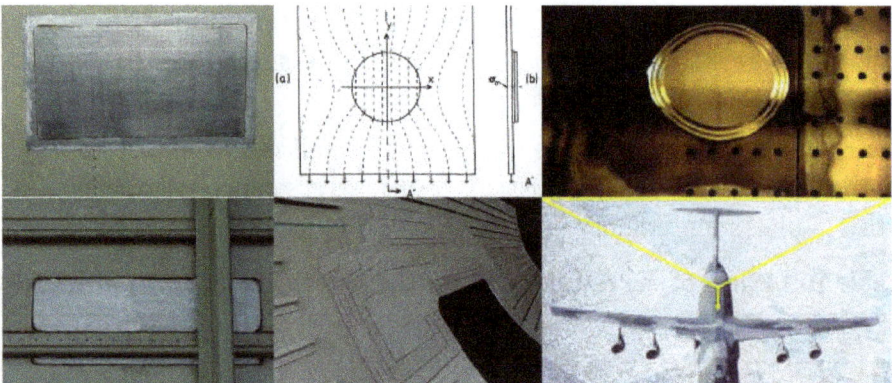

Figure 10.11: Reinforcements using doublers or patch repairs may induce stress concentrations. (TU Delft, N.D.)

10.4 FATIGUE

10.4.1 Definition

It has been mentioned in the introduction to this chapter that fatigue as phenomenon has been well understood for metallic materials and structure, see Schijve (2001). However, despite this knowledge and understanding, there seems to be misperception with respect to the phenomenon of fatigue when considering different material types, like for example fibre reinforced composites. The misperception is illustrated with the statement often put forward, that composites do not suffer from fatigue.

If that statement is made, it may refer to two cases:

- When cyclic loading composites in tension, similarly to common loading of metallic materials, composites hardly show evidence of damage initiation and growth, where fatigue cracks may initiate and propagate in metals.

- When designing a composite structure, addressing the static strength requirements usually is sufficient to cover fatigue.

The first aspect may be true, but in return there may be repetitive load cases that do not cause any fatigue damage in metallic materials, that do initiate and propagate damage in composites. As example, certain compressive load cycles severely damage composite structures when applied many times, which do not cause any fatigue damage in metallic materials.

The second aspect, relates partly to the discussion of the stress concentration factor; linear elastic material is very sensitive to notches, which is not the case for metallic materials. Therefore metallic materials can be more efficiently loaded up to their static strength. In addition, reduction due to environmental impact (see chapter 2), further restricts the static allowables of a composite design. When these restrictions are accounted for, the corresponding load spectra do not cause fatigue issues in the structure.

However, the statement and the above mentioned considerations may never refer to the conclusion that composites do not face damage phenomena under any repetitive or cyclic load spectra. In general, it can be stated that all engineering materials in one way or another suffer from fatigue. Only the phenomena appear to be different leading to different failure mechanics under different loading conditions.

To do justice to the generic aspect of fatigue, fatigue can therefore be defined as:

Damage phenomenon induced by a large number of load cycles below the ultimate strength of material or structure causing permanent deterioration of material or structure resulting in a reduction in load bearing capability.

For aeronautical structures, the load cycles mentioned in this definition are often related to number of flights, ranging from approximately constant amplitude (fuselage pressurization every flight) to arbitrary spectrum loading (wing loading during take- off, landing, and turbulence). This is illustrated with an arbitrary flight load spectrum shown in Figure 10.12. In this example the maximum load in the whole spectrum is called nominal or limit load. This is the load case that is considered to occur only once in a lifetime. At this load failure may not occur. Ultimate load, used for strength justification is usually calculated as 1.5 times the nominal or limit load.

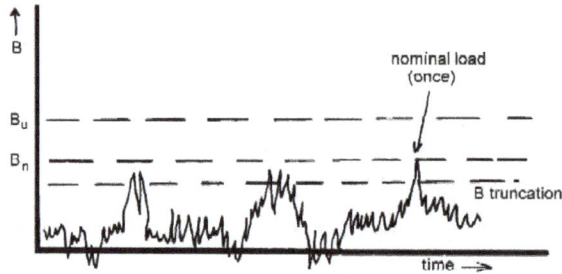

Figure 10.12: Illustration of an arbitrary flight load spectra. (TU Delft, N.D.)

10.4.2 Fatigue assessment

It was noticed by the researchers investigating fatigue after the railroad failures that fatigue appears to be predominantly influenced by the amplitude of the load cycles. Wohler, who performed an extensive research at the time on fatigue loading of train axles, concluded that the

"Material can be induced to fail by many repetitions of stresses, all of which are lowerthan the static strength. The stress amplitudes are decisive for the destruction of thecohesion of the material. The maximum stress is of influence only in so far as the higher it is, the lower are the stress amplitudes which lead to failure"

For this reason, the fatigue characteristics are traditionally represented by so called Wohler curves, or S-N curves, see for example Figure 10.13.

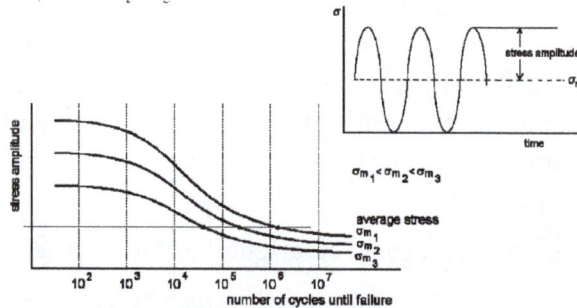

Figure 10.13 Illustration of a Wohler curve (stress amplitude is decisive). (TU Delft, N.D.)

One should carefully pay attention to the S-N curve as illustrated in Figure 10.13. In this figure, three curves are given for three different mean stress levels. It is illustrated that the lower the mean stress the higher the curve. This relates to the conclusion of Wohler; the stress amplitude is decisive and thus plotted on the vertical axis, but there is an effect of maximum stress. The lower the mean stress forgiven stress amplitude, the lower the maximum stress. This will increase the fatigue life (equivalent to shifting the curve to the right), which is equivalent to shifting the curve upward.

To understand this, one has to consider the stress cycle. The relation between the amplitude stress σ_a and the mean stress σ_{mean} on the one hand, and the minimumstress σ_{min} and maximum stress σ_{max} on the other hand is given by:

$$\sigma_a = \frac{\sigma_{max} - \sigma_{min}}{2} \; ; \; \sigma_{mean} = \frac{\sigma_{max} + \sigma_{min}}{2} \tag{10.6}$$

Note that the assessment of fatigue using S-N curves is solely based on evaluating fatigue under constant amplitude loading, i.e. all load cycles have the same mean stress and stress amplitude. As earlier mentioned, except for pressurization of fuselages all flight load spectra are random of nature and thus to be considered as variable amplitude loading.

In general, two asymptotes can be identified in the Wohler curve

◉ Upper asymptote, related to the maximum stress reaching the ultimate strength

◉ Lower asymptote, defined as fatigue limit

The lower asymptote is an important parameter, when evaluating fatigue for specificmaterials or structures. Especially, when high cycle fatigue is considered, i.e. fatigue with a very large number of cycles (order 10^5 or higher). The fatigue limit is defined as the stress amplitude (usually given for a mean stress of $\sigma_{mean} = 0$), below whichfatigue failure do not occur. This does not mean that nucleation of microscopic fatigue damage does not occur, but that even if such damage does occur that it will never propagate to macroscopic lengths causing failure.

One way to design against fatigue is to limit the operational load spectra to stress amplitudes below the fatigue limit. This approach formed the basis for the Safe-Life concept discussed in chapter 9.

10.4.3 Variable amplitude loading

The assessment of flight load spectra, or variable amplitude fatigue loading, is in general performed using the constant amplitude data as represented in Wohler curves, see Figure 10.13. The most well-known method widely applied to do approximate the variable amplitude behaviour (and also well known to be very inaccurate) is the Miner rule. This rule is a linear damage accumulation rule that sums the damage fractions of each individual load cycle in the load spectrum until the total damage fraction reaches the value 1. The material or structure is then considered failed.

The damage fraction of for example the ith load cycle in the spectrum is calculated with

$$D_t = \frac{1}{N_i}$$

(10.7)

where N_i is the fatigue life determined with the constant amplitude S-N data, see Figure 10.13. If the load spectrum contains more of the same load cycle, this relationis written as:

$$D_t = \frac{n_i}{N_i}$$

(10.8)

Failure of the component then will be assumed to occur at:

$$\sum_i^n \frac{n_i}{N_i} = 1$$

(10.9)

Again, one should note that this simple rule ignores several aspects that play a role in the fatigue phenomenon. As a consequence, this rule is often inaccurate. The engineering approach to account for that is to calibrate the rule to specific materials and load spectra. This calibration, or more correct empirical correction, may lead to values smaller or larger than 1.

10.4.4 Notched materials and structures

The above discussion has been presented assuming there are no notches present in the material or structure. The fatigue limit mentioned in section 10.4.2 is then considered to be a material parameter. However, following from the discussion of stress concentrations at the beginning of this chapter, it is obvious that fatigue may occur sooner at a notch where a peak stress is present.

The effect of a stress concentration, described by the factor K_t, on the fatigue behaviour can be illustrated with the S-N curve as given in Figure 10.14. The upper asymptote, related to K_t, is not affected for metallic materials, because of the plasticity redistributing the stress, see Figure 10.5. Only a small, often negligible, reduction will occur. Be aware that this is not the case for linear elastic materials, likecomposite materials. For these materials the upper limit will reduce by a factor .

For both metallic and composite materials, there is a significant effect on the fatiguelimit. Because of the higher peak stress at the notch, fatigue damage may initiate sooner than would be expected based on the undisturbed stress field data. In general, the fatigue limit reduces by a factor K_t for notched structures.

For small notches and large stress concentration factors K_t, this reduction may be less severe for metallic materials. Again the plasticity provides an advantage over linear elastic materials. High values for K_t implies theoretically high values for the peak stress. However, similar to the discussion on Figure 10.5, the peak stress is levelled off beyond the yield strength of the material. As a consequence, the real peak stress is lower.

Following from the definition given by equation (10.3) , this implies a lower stress concentration. The actual stress concentration due to plasticity is therefore described by a notch factor K_f, which is lower than the stress concentration factor K_t. Again, note that this only the case for ductile (thus elastic-plastic) materials and not for linear elastic materials, like composites. Composites are considered to be very notchsensitive.

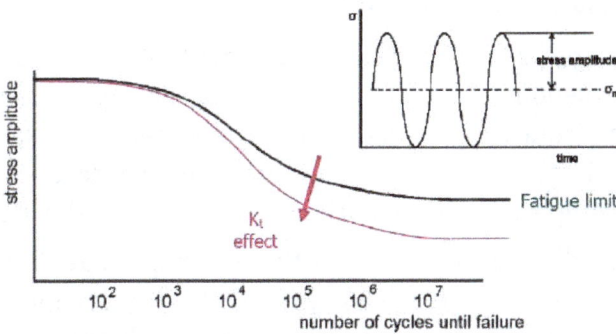

Figure 10.14: Effect of stress concentration on the constant amplitude fatigue life. (TU Delft, N.D.)

10.4.5 Designing against fatigue

The fatigue behaviour represented by the S-N curves is considered up until failure. It was already mentioned that below the fatigue limit, microscopic damage may occur but will not lead to failure. The fatigue life governed by the S-N curve, consists for 80-90% of formation of a fatigue damage of about few millimetres (nucleation of microscopic damage up to macroscopic lengths). Only a few percent of the life consist of a crack that propagates to a larger length causing final failure.

The Safe-Life concept implies designing against fatigue using fatigue life data (S- N curves). To justify such design, large safety factors are required with respect to available fatigue data. The experimental life (proven by tests) must be about 3 to 4 times the life that the aircraft is certified for. This is related to the fact that the experimental data shows considerable scatter, especially at lower stress amplitude cycles.

Although the S-N curves are presented like in Figure 10.13 and Figure 10.14, one has to be aware that these curves represent average or minimum data. Performing individual tests near the fatigue limit for example, may result in fatigue lives that deviate by a factor 10 from the curve.

Because of the large margin of safety applied using the Safe-Life concept, the thickness of fatigue critical parts is often significant larger than required for static strength. Especially, for metallic materials this is the case.

Designing against fatigue then aims to increase fatigue lives or to increase allowable fatigue stresses for given operational life. This implies lowering stress concentrations and avoidance of damage initiators in the design, but also selection of materials thatare less fatigue sensitive (higher fatigue limits). However, this can be achieved also by applying a damage tolerance philosophy rather than a Safe-Life concept.

40.4.6 Some fatigue characteristics

It has been mentioned a few times in this chapter that fatigue is a generic damage formation phenomenon related to cyclic loading. The appearance of the damage may be different from one material type to another. For example, in metallic materials, thecyclic (tensile) loading may initiate and propagate cracks, whereas in fibre reinforcedcomposites, delaminations and transverse shear cracks may occur.

The type of loading for formation of these damage types is often different. As said, cracks in metals are often driven by tensile load cycles applied in plane of the material. The delaminations in composites are often induced by cyclic compression, shear or bending loads. Looking at these materials on a microscopic level, one may see some qualitative commonalities that play a role in the nucleation of damage. In metals and alloys, often the inclusions in the material cause stress concentrations at microscopic level,because the mechanical properties of inclusion and surrounding material are different, see Figure 10.15. However, this difference in mechanical properties is also present for longitudinal fibres and matrix in composites. Qualitatively, both cases imply stress concentrations that may case tensile cracks in metals and transverse shear cracks and delaminations in composites.

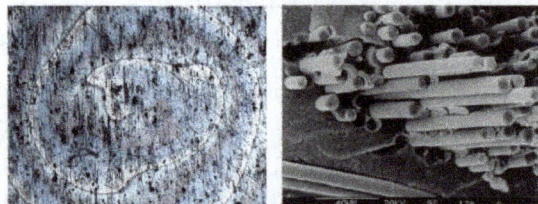

Figure 10.15 Qualitative comparison between metal alloys containing inclusions (left) and fibre reinforced composites (right); in both cases stress concentrations are present at micro level. Derivative from left: Antiproton (2017), CC0 and right: MT Aerospace AG, (2006), CC-BY-SA3.0.

Despite the qualitative microscopic correlation in Figure 10.15, fatigue appears to be different in metals and composites. The fatigue damage of both cases in this figure are nucleated by different loads on a macroscopic level; metal are most sensitive to tension-tension loading, while composites are most sensitive to compression- tension or compression-bending loading. Some illustrations of typical fatigue failure mechanisms on a macroscopic level are illustrated in Figure 10.16.

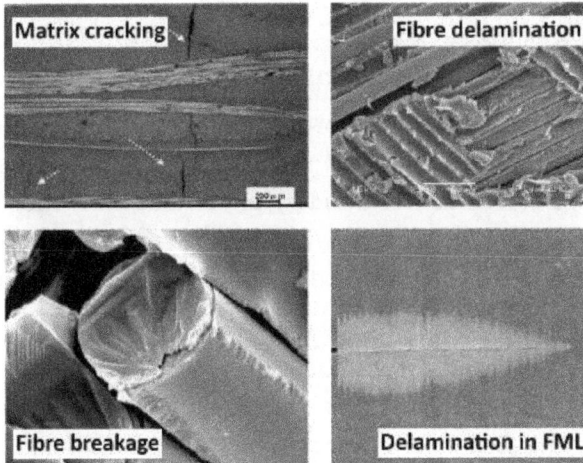

Figure 10.16: Illustrations of typical fatigue failure mechanisms in fibre reinforced composites and fibre metal laminates. Derivative from top right: Carnevale, (2014), Reprinted with Permission.; top right: Alderliesten, (2011), 11-16-b.jpg, Own Work; bottom left: Plymouth Electron Microscopy Centre, (n.d.), Copyright PEMC. Reprinted with permission; bottom right: Plymouth Electron Microscopy Centre, (n.d.), Copyright PEMC. Reprinted with permission.

10.5 DAMAGE TOLERANCE

10.5.1 Definition

There are different definitions for damage tolerance. Following from the discussion in chapter 9, damage tolerance represent a design philosophy where structural integrity is assured even in presence of potential damages by a combination of built-in design features and inspection and maintenance procedures. This definition is very relevant for the aeronautical field of structural design.

However, sometimes a more strict or limited definition is used that relates to the former definition, but omits the aspect of possible inspection and maintenance. This definition relates to the material's response to potential damage cases. Traditionally, this definition included fatigue cracks and environmental damages, but also impact damages. Interestingly enough, with the introduction of composites in primary aircraft structures, people seem to further limit this definition to impact damages alone. Thisis probably related to the fact that composites are considered not to suffer from fatigue, which has been discussed before.

However, it is recommended to consider all relevant damage scenarios even when addressing damage tolerance as a material aspect only. This aspect is considered in the definition of residual strength, which considers the remaining load bearing capability of material or structure in presence of potential damage. In the last section of this chapter, this aspect will be briefly addressed for a material or structurecontaining a fatigue crack.

10.5.2 Limitations of stress concentration concept

The stress concentration factor K_t captures the effect of disturbance in the stress field due to the presence of a notch. However, when addressing the residual strength(or damage tolerance) of a material or structure, there is a limitation to the concept. This can be illustrated with the description of equation (10.5). The stress concentration is directly related not only to the size of notch, but also the radius at the critical location.

Consider that a fatigue crack has initiated at the edge of a hole and propagated a few millimetres, he question what the residual strength is cannot be answered using the stress concentration factor. Because a crack has a sharp tip, which may be considered to have a infinitely small radius. With equation (10.5) this implies that thestress concentration factor becomes infinite, indifferent from the size of the crack. However, it does not require a lot of fantasy to understand that a structure with a small crack is capable to carry a higher load than the same structure containing a long crack. In other words, to assess the residual strength another parameter is needed than the stress concentration factor K_t. What can be learned from equation (10.5) is that apparently, a crack with a sharp tip induces a singularity in the stress field right at the tip. To mathematically derive a relevant parameter describing this stress singularity first a crack is considered in a sheet with infinite dimensions. This case is considered for similar reasons as explained for the stress concentration factor: the boundary conditions are exactly known.

The stress field including the singularity can be mathematically derived for this case.Although not necessary to reproduce in this course, for illustration the stress field is given here

$$\sigma_x = \frac{S\sqrt{\pi a}}{\sqrt{2\pi r}} \cos\frac{\theta}{2}(1 - \sin\frac{\theta}{2}\sin\frac{3\theta}{2}) - S$$

$$\sigma_y = \frac{S\sqrt{\pi a}}{\sqrt{2\pi r}} \cos\frac{\theta}{2}(1 + \sin\frac{\theta}{2}\sin\frac{3\theta}{2})$$

$$\tau_{xy} = \frac{S\sqrt{\pi a}}{\sqrt{2\pi r}} \cos\frac{\theta}{2}\sin\frac{\theta}{2}\sin\frac{3\theta}{2}$$

$$(10.10)$$

It can be observed that these equations have a component in the equation that describes the location measured with respect to the crack tip. The other part in the equations has a similar appearance for each equation. It is this part of the equation that is taken as parameter, indicated by K.

$$K = S\sqrt{\pi a}$$

(10.11)

This parameter is called the stress intensity factor (not to be confused with stress concentration factor). In this relation the crack length is evidently related to the dimensions of the crack (which was no longer described with K_t). In addition, the applied stress levels have an influence on the stress intensity.

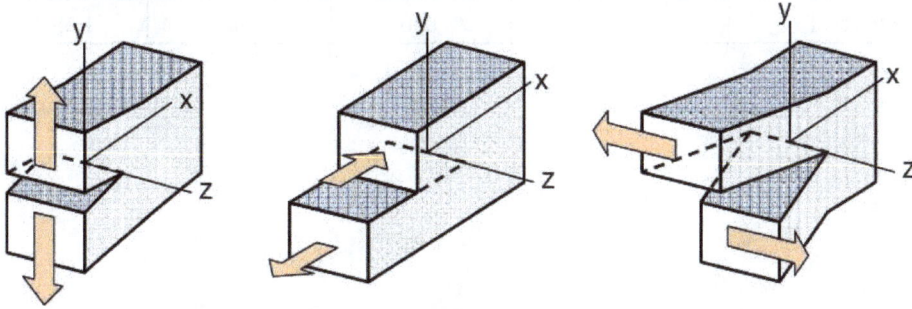

Figure 10.17: Three different opening modes of a crack; tensile mode I, shear mode II, and transverse shear mode III. (Alderliesten, 2011)

When considering cracks, different opening modes can be identified, see for exampleFigure 10.17. These opening modes imply different behaviour. Obviously, fatigue cracks in metallic materials and structures under cyclic tensile loading predominantly focus on the tensile mode I.

The stress intensity factor K for a crack opening in mode I, is indicated with (subscript denotes the opening mode). The residual strength of a material containing a crack is evidently then related to the critical value of this K_I. This critical parameter,K_{Ic} is called the fracture toughness, previously discussed in chapter 1. This fracture toughness is considered a material specific material. Materials with high fracture toughness are clearly capable to sustain high loads in presence of cracks, or considerable loads in presence of very long cracks, see equation (10.11).

Thus, if the fracture toughness is given by:

$$K_{Ic} = S_{crit}\sqrt{\pi a_{crit}}$$

(10.12)

The sensitivity of materials to cracks can be illustrated with Table 10.2, assuming that the critical stress is related to the material's yield strength by $\sigma_{Crit} = \sigma_{0.2}/2$.From the comparison it can be observed that despite steel having the highest fracture toughness of these four materials, the critical crack length is the smallest. Evidently, the influence of ductility plays an important role here. Ductile alloys, having low yield strength, are usually less sensitive for cracks.

Table 10.2 Comparison between critical cracks lengths of some aerospace alloys

Alloy	$\sigma_{0.2}$ [MPa]	σ_{crit} [MPa]	K_{Ic} [MPa m1/2]	a_{crit} [mm]
2024-T3	360	180	40	15.7
7075-T6	470	235	27	4.2
Ti-6Al-4V	1020	510	50	3.1
4340 steel	1660	830	58	1.55

CHAPTER-11

STRUCTURAL JOINTS

11.1 INTRODUCTION

In previous chapters various aircraft and spacecraft structures have been discussed briefly with respect to their shape and functions. The analysis of general load paths and the way loads locally are translated into have been identified.

However, all these structures are constructed from parts that each often are build-up from smaller elements. All these elements and parts are jointed together to form the complete structure. These structural elements have been identified previously; sheets, stringers, ribs, frames, web plates, girders, clips, etc.

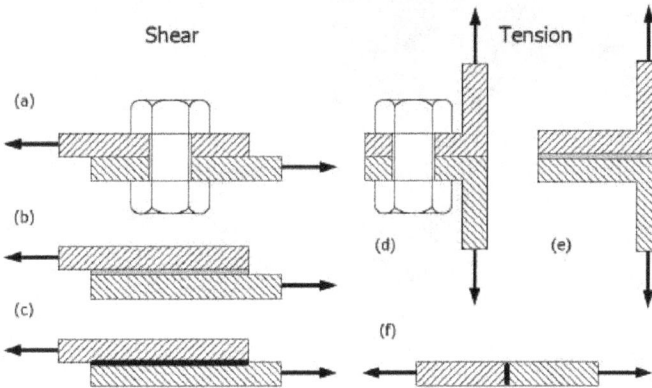

Figure 11.1: Two load transfer modes to classify the three major joining techniques in aerospace; shear (a-c) and tension (d-f). (TU Delft, N.D.)

There are quite a few joining techniques currently applied in engineering; only few of them are considered applicable in aerospace engineering. The most commonly applied joining techniques in aerospace are:

- ⊙ Mechanically fastening (includes riveting and bolting)
- ⊙ Welding
- ⊙ Adhesive bonding

The three major joining techniques can be classified by the loading mode of the joint;the joint may either transfer load in tension, or in shear, see Figure 11.1. Although in principle each joining technique could be designed as either tensile or shear joint, thetensile load transfer is avoided for adhesive bonding.

This chapter provides general characteristics of each of these joining techniques. These characteristics may be used to identify the advantages and disadvantages of one joining technique over the other. Here, care must be taken, because the advantages and disadvantages are relative; what may be considered in one structure as advantage may imply a disadvantage in another structure. These considerations will be discussed in the next chapter.

11.2 MECHANICALLY FASTENED JOINTS

Different mechanical fasteners are used in engineering. The four major types are illustrated in Figure 11.2. The traditional nails often applied in wooden constructions,are considered inapplicable in aerospace, and will therefore not be discussed here. The three remaining fastener types are threaded fasteners, blind fasteners and rivets. Threaded fasteners are commonly referred to as bolts.

Figure 11.2: Illustration of the four major fastener types: threaded fasteners (top left), rivets (top right), blind fasteners (bottom left), and nails (bottom right). Derivative from top left: Alexas_Photos, (2016), Public Domain; top right: Saunders-Smits, (2018), 11-2-b.jpg. Own Work; bottom left: Cdang, (2010), Public Domain; bottom right: InspiredImages, (2015), CC0.

The lap joint and butt joint are typical shear joints applied in thin walled structures. The thickness step in the lap joint is oriented longitudinally in the fuselage structure, which implies no aerodynamic problem. The circumferential shear joint have to be smooth for aerodynamic reasons. Therefore, the butt jointis applied.

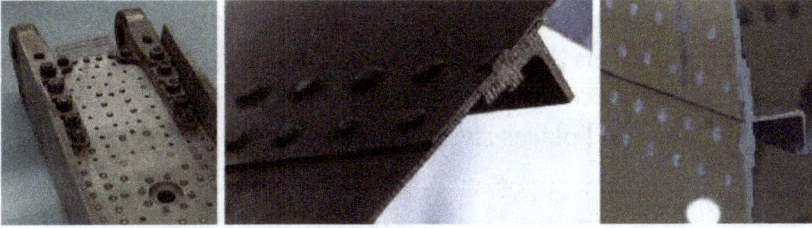

Figure 11.3: Wing spar connection (left), fuselage lap joint (centre) and fuselage butt joint (right). (TU Delft, N.D.)

Figure 11.4 illustrates typical tensile joints in aeronautical structures. The load into the horizontal stabilizer is introduced through the lower skin sheets into the ribs and spars. The joint exhibits a lot of tensile joints through several additional skin doublers. The channel fitting is a typical design feature to introduce tensile load by the bolt gradually into the skin sheets. The gradual reduction of the fitting's cross section by tapering the flanges implies stiffness reduction. The rivets in the fitting are typical shear joints.

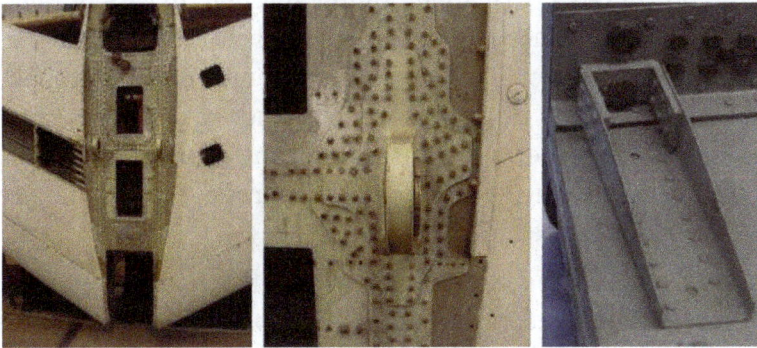

Figure 11.3: Wing spar connection (left), fuselage lap joint (centre) and fuselage butt joint (right). (TU Delft, N.D.)

11.2.1 Rivets

Riveting is generally applicable for joining sheet material in which the joint transfer the load from one sheet to the other in shear. Because the rivet is being forced into its final shape during installation, access from both sides is needed. In case access can be provided at only one side, blind rivets are used.

Riveting is a reliable joining method because it has been applied for many decades (lot of experience), but also because the joint can be well inspected and repaired. In case a riveted joint must be repaired rivets can be drilled out and replaced by a slightly larger rivet.

Riveting is often considered for its low cost; both the rivets and the installation per unit are fairly cheap. For rivet installation, only pneumatic hammers or rivet guns are necessary tooling to apply the riveting force, while bucking bars are needed to counteract the applied force. An alternative method for riveting with pneumatic hammers is the application of squeezing. The rivet is then pressed into its final shape by a force controlled machine. This method for rivet installation provides better quality and places rivets in a more reproducible manner.

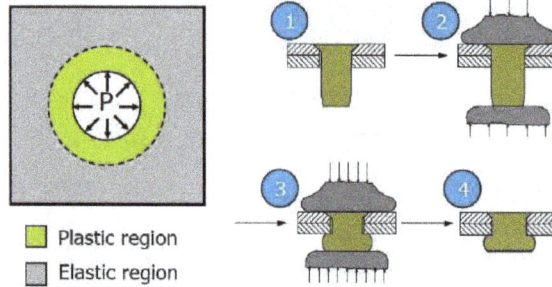

Figure 11.5: Illustration of the riveting principle: rivet placement (1), riveting squeeze force (2), rivet deformation (3) and final result (4).(TU Delft, N.D.)

Riveting is considered a permanent joining method. In case of repair the rivet could be drilled out and replaced by an oversized rivet (rivet with larger diameter), but this can only be done very few times. Each removal implies application of a larger rivet, which can only be done a limited amount of times. Another reason why riveting is considered permanent is that during operational loading the rivet cannot vibrate loose. Although the residual compressive stress may reduce over time, the rivet remains fixated in the hole.

Because the rivet is forced into the hole, it will fill the hole completely. During riveting, the shaft of the rivet will expand (see chapter 1), applying radial pressure to the sheet edges. As a consequence, the stress concentration factor induced by the holes, see chapter 10, is reduced by radial expansion of the rivet. The riveting principle is illustrated in Figure 11.5.

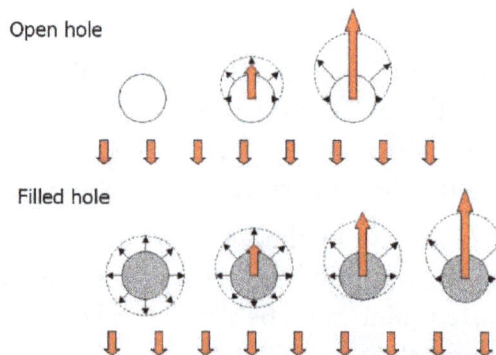

Figure 11.6: Difference in stress concentration due to load transfer induced by bearing pressure only (open hole) and load transfer by bearing and radial compression (filled hole). (TU Delft, N.D.)

The remaining residual stress is considered favourable for the load transfer, and therefore also for the fatigue properties of the riveted joint. The reason can be explained with the two cases illustrated in Figure 11.6. The open hole configuration could represent a bolted joint in which the bolt is placed free from the sheet edges, often denoted as clearance fit. The load applied by the bolt to the sheet edges is called bearing pressure. This bearing load must be in equilibrium with far field stresses further away in the sheet. As a consequence, all stress is concentrated near the hole edges, implicating a large stress concentration.

The filled hole represents a rivet that has been squeezed into the hole applying radial expansion to the sheet edges. Subsequent loading of the joint, would not only apply bearing pressure to the sheet edges, but would also reduce the radial compressive forces at the back side of the hole. As a consequence, the load around the hole edges to be in equilibrium with the far field stresses in the sheet is significantly less. In other words, riveting reduces the stress concentration factor K_t at the hole edge by radial expansion of the rivet shaft.

11.2.2 Threaded fasteners

The threaded fasteners have a wider range of application. First of all, they can be applied both in shear joints and tension joints, and in joints that comprise a combination of both loading modes. In addition, bolts can be applied not only to sheet material, but also to a wider variety of structural components.

A general characteristic is that threaded fasteners, often called bolt, allow disassembly after being assembled. In other words, they are not permanent but can be removed and re-installed again. Where a rivet after being drilled out is discarded and replaced, bolts can be used again.

Another characteristic of threaded fasteners is the wider range of materials that can be selected to manufacture the bolt. Rivets must be deformable during installation, which limits the type of materials from which it can be manufactured. For threaded fasteners, this limitation is not present, and as a result there is more freedom to select the appropriate bolt material.

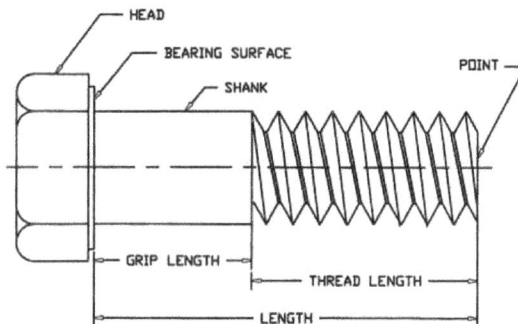

Figure 11.7 Definition of the bolt dimensions. (TU Delft, N.D.)

Therefore high strength joints are often bolted joints. The strength of bolts can be tailored by selecting the right material, the appropriate heat treatments, and convenient case hardening technique. The dimensions of the bolt can be chosen depending on the to-be-jointed components. An illustration of dimension definition isgiven in Figure 11.7.

11.2.3 Load transfer mechanisms

In mechanically fastened joints multiple load transfer mechanisms act together to provide the load path through the joint. Figure 11.8 illustrates the primary and secondary loads that act in a single sided shear joint.

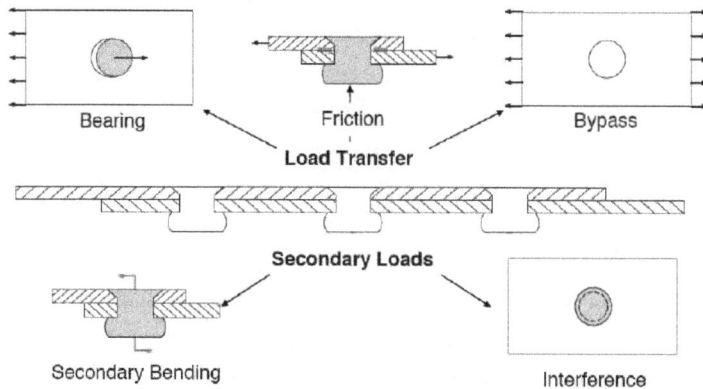

Figure 11.8 Illustration of the primary and secondary loads in mechanically fastened joints. (TU Delft, N.D.)

To evaluate a mechanically fastened joint, the individual load paths must be identified. In a single row joint, as illustrated in Figure 11.9 all load is transferred form one sheet to another by either bearing pressure on the sheet edges, or by friction between the sheets. The magnitude of friction depends on the clamping force that is provided by the bolts.

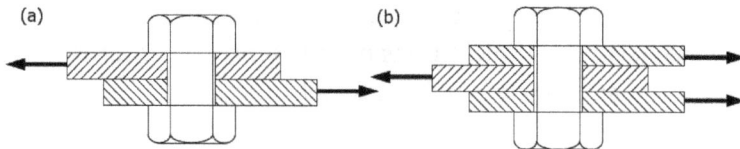

Figure 11.9 Illustration of a single sided shear joint (a) and double sided shear joint (b). (TU Delft, N.D.)

In case of a multiple row joint, for example three rows as illustrated in Figure 11.8, allload is distributed over the three rows. This implies that the load in the upper sheet in Figure 11.8 is partly transferred to the lower sheet at the first rivet row, by bearing and friction, while the remainder of the load continues through the upper sheet to thesecond and third rivet row. This part of the load must then go around the rivets in thefirst row, which is called by-pass load.

Secondary loads to be considered are the interference loads, which are created by rivets that expanded in the holes during riveting, or by bolts with an oversized shank diameter that have been forced into an undersized hole. These loads add to the bearing pressure as illustrated in Figure 11.6.

The main difference between the two shear joints in Figure 11.9 relates to symmetry.The single sided shear joint is an asymmetric joint, whereas the double-sided joint can be a symmetric joint. In case of a symmetric joint, equal load is transferred over both interfaces, which implies that the deformation of the joint remains symmetric. In an asymmetric joint, the load transfers through one interface, creating an asymmetric step in the load path. This load step induces an asymmetric deformation in addition to the longitudinal deformation induced by the applied load, which is called secondary bending, see Figure 11.10.

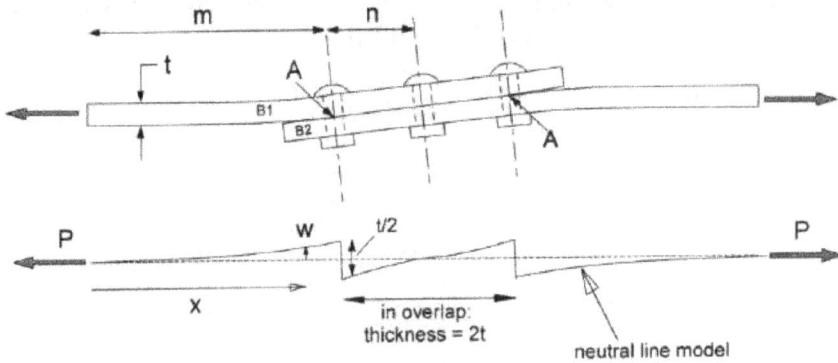

Figure 11.10 Illustration of secondary bending in a single sided shear joint; the deformation relates to the load step correlated to the neutral line through the joint. (TU Delft, N.D.)

Consider load transferred by bearing in the joint in Figure 11.11. The bearing *pressure is defined as:*

$$\varepsilon = \frac{\Delta L}{L}$$

where F is the load transferred by bearing, D is the diameter of the hole and t the sheet thickness. The bearing pressure, or bearing stress, has the units [N/ mm2] or [MPa]. This strength is considered characteristic for sheet materials, as it describes when sheets will deform and fail under bearing pressure. In a symmetric joint, in absence of secondary bending, the bearing pressure is treated as a homogeneous stress distribution through the thickness of the sheet. Secondary bending implies a variation of stresses through the thickness of the sheet as illustrated in Figure 11.11. The superposition of two stress distributions implies that maximum stress is reached at one side first.

Considering fatigue, it also implies that at one side the stress cycles are greater Introduction to Aerospace Structures and Materials 229 and higher than at the other side. The detrimental influence of secondary bending on the fatigue life is illustrated with the S-N curves in Figure 11.12.

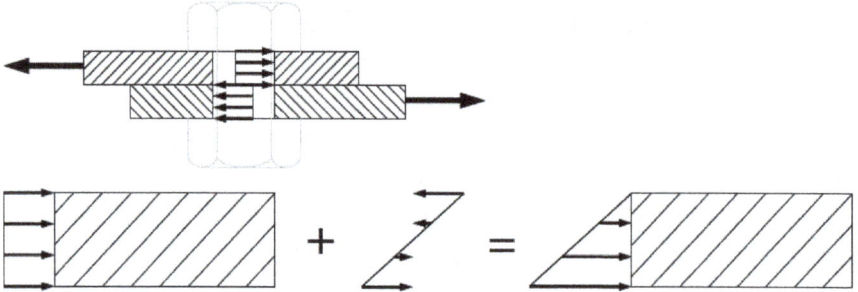

Figure 11.11: Peak stresses due to superposition of bearing stress and bending stress. (TU Delft, N.D.)

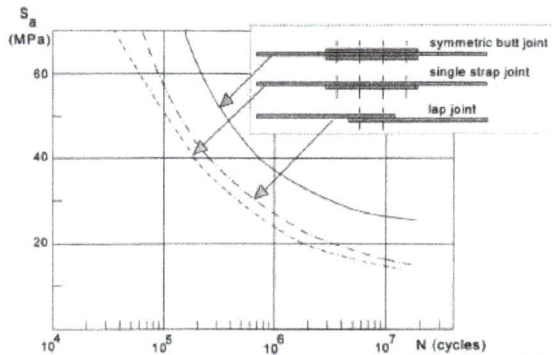

Figure 11.12: Effect of secondary bending on the fatigue performance of riveted joints. (Schijve, 2010, Copyright Schijve, used with permission)

11.2.4 Failure modes

The strength of mechanically fastened joints relates to both the fastener (type and material) and the jointed sheets. The joint can fail in different ways, which are identified as the joint failure modes, illustrated in Figure 11.13.

One failure mode relates to rivet failure; the shear load in the rivet exceeds the ultimate shear strength of the rivet material, causing shear failure of the rivet. In general, this failure mode should be avoided in aeronautical structures. Once the load on a rivetedjoint causes rivet failure, it can be expected that subsequent redistribution of load over the other rivets will cause failure of these rivets as well. As a consequence, the complete riveted joint may fail catastrophically.

The shear failure load on a rivet can be calculated for a single shear joint (see Figure 11.9) with

$$F_{ult} = \frac{\pi}{4}D^2\tau_f$$

$$(11.1)$$

Where D is the diameter and τ_f is the failure shear strength of the material. For a double sided shear joint the load will be twice as high, because the load transfer takes place at two distinct interfaces.

The preferred failure modes relate to sheet failure. The three sheet failure modes in riveted joints are net-section failure, bearing failure and shear-out (or tear out) failure, see Figure 11.13. The net-section failure relates to load exceeding the ultimate tensile strength of the sheet in the cross-section between the rivets. For a single rivet that can be written as

$$F_{ult} = (W - D)t\sigma_f$$

$$(11.2)$$

Where W is the width, t the sheet thickness, and σ_f the failure strength of the material.

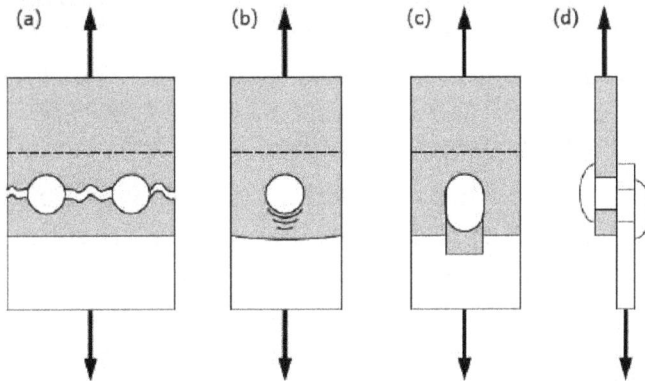

Figure 11.13: Four major failure modes for mechanically fastened joints. (Alderliesten, 2011)

The second failure mode related to sheet failure is the bearing strength failure. The bearing strength failure can be calculated with (see example on the effect of secondary bending).

$$F_{ult} = Dtp_b$$

$$(11.3)$$

However, considering bearing strength in riveted joints, often the permanent deformation is considered for determining the nominal allowable load. Based on experience the maximum permanent deformation allowed is set to be 2% ovalisation of the hole, which is equal to 0.02D. The bearing pressure at which this deformation occurs is defined as n % . Thus the nominal load should satisfy

$$F_{nom} \leq Dtp_{2\%}$$

$$(11.4)$$

The ultimate load is then obtained by multiplying both sides of the equation with the safety factor of 1.5, which yields

$$F_{ult} = 1.5Dtp_{2\%}$$

$$(11.5)$$

The last sheet failure mode is shear out of the sheet under the bearing pressure. This failure mode relates to the edge distance, i.e. the amount of material between the rivet and the edge of the sheet, see Figure 11.14.

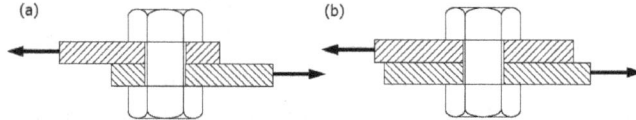

Figure 11.14: Small edge distance (a) will cause shear out failure, which can be avoided with greater edge distances (b). (TU Delft, N.D.)

Of course, the maximum load that a mechanically fastened joint can sustain is the smallest of the four failure loads.

11.3   MECHANICALLY FASTENING IN COMPOSITES

Most of the discussion in the previous section applies to composite structures. However, several additional aspects should be considered. It was explained in chapter 1 that the orthotropic nature of composite panels implies that the strength depends on the orientation of the fibres within the panel.

This can be explained considering a single ply of fibre reinforced composite, illustrated in Figure 11.15. The longitudinal plies are strong in fibre direction, but the shear strength of the matrix material between the fibre is insufficient to resist the shear- out failure. Transverse plies have insufficient resistance against the bearing pressure, most likely causing net-section failure.

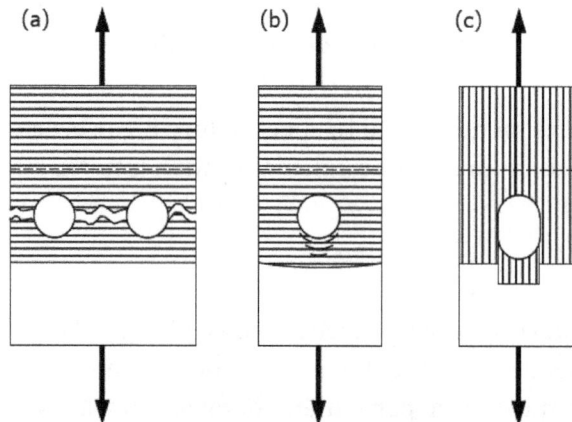

Figure 11.15: Relation between fibre orientation and failure mode; longitudinal plies are weak against shear-out, while transverse plies are weak against net-section failure and bearing. (Alderliesten, 2011)

This implies that the composite laminates that are jointed together require plies oriented in several directions to create sufficient resistance against either of the three failure modes. Because the shear-out failure mode is induced by shear stresses between the fibres, see Figure 11.15 (c), often fibre layers are added oriented in ±45° orientation, because of their shear strength contribution.

To create sufficient bearing strength in composite panels, sufficient layers in 0°, 90°, and ±45° are needed. This implies that the thickness near the edges of composite panels often must increase to create sufficient quasi-isotropic thickness as is illustrated in Figure 11.16. As alternative, to limit the thickness increase often metallic inserts are added, because of their isotropic nature. Rather than applying multiple fibre layers in a quasi-isotropic lay-up (see definition in chapter 1), a single isotropic metal layer will suffice.

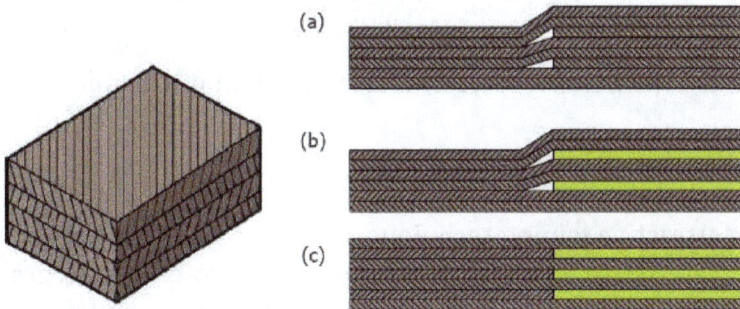

Figure 11.16: Illustration of increasing bearing strength in composite panels near the edge; increasing thickness (a), adding metallic inserts (b), and replacing layers with metallic inserts (c). (Alderliesten, 2011)

11.4   MECHANICALLY FASTENING IN SANDWICHCOMPOSITES

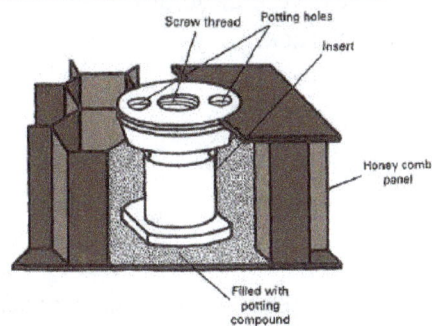

Figure 11.17: Application of inserts in the sandwich core to create suficient bearing strength.(TU Delft, N.D.)

The application of mechanically fastening in composite sandwich panels is somewhat more complex compared to monolithic composite panels, discussed in the previous section. The complication is related to the characteristics of facing and core material; where the facings transfer all normal loads, the core only transfers the shear load. This means that the mechanical fastener in a sandwich panel has to transfer its load by bearing mainly to the two face sheets of the sandwich, as is illustrated in Figure 11.17 and 11.18

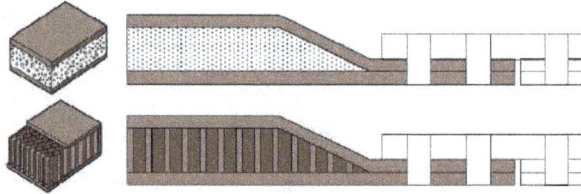

Figure 11.18: Reduction of sandwich panel to monolithic composite edges to create suficient bearing strength. (Alderliesten, 2011)

11.5 WELDED JOINTS

In this section the welding of joints is discussed for both metallic and composite structures.

14.5.1 Metallic structures

Welding is an attractive joining technique, because a good weld is almost as good asthe parent material. In general, welding provides a so-called integral structure. After welding, the structural component can no longer identified as separate components.

For aeronautical purposes where fatigue and damage tolerance is of importance, this has a drawback. As illustrated in Figure 11.19, a crack propagating in the skin sheet, will sense no natural barrier against growth into the welded stringer, whereas the riveted and adhesive bonded joint provides a natural barrier for the propagating crack.

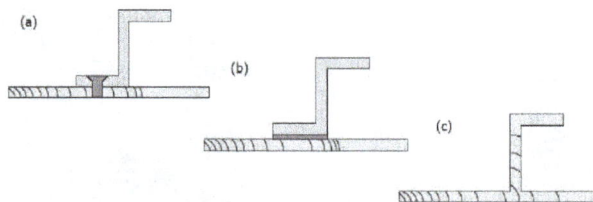

Figure 11.19: Comparison between mechanically fastened, bonded and integral structures; integral structures do not provide barriers against cracking. (Alderliesten, 2011)

There are two major welding techniques applied in aerospace structures:

- ◉ Laser beam welding
- ◉ Friction stir welding

In both cases heat is added to the material, but the principle differs significantly. In case of laser beam welding, the laser beam is the heat source that very locally heats the material to melting temperature levels. Because of the welding rate and the local rapid rate of heating and cooling, only a small area is affected by the heat treatments, the so-called heat affected zone.

Friction stir welding is a different process. It is not based on heating material to melting temperature, but it is a process in which the material remains in its solid state. A cylindrical tool with shoulder is rotated and pressed with great force into the sheets. The friction between tool and sheet material generates sufficient amount of heat to soften the material and to allow mixing of the plasticized flow. Because of the limitedamount of heat and the temperature levels, a smaller area of the weld is affected by the heat treatment. In addition, friction stir welding does not require addition of weld material, only the parent sheet materials are mixed during the process.

Whereas laser beam welding can only be applied to certain aluminium alloys because of the applied heat treatments, friction stir welding can be applied to more alloys including the high strength aeronautical alloys. It even has been proven that dissimilar alloys can be welded with friction stir welding.

11.5.2 Composite structures

In composite structures welding can only be performed with thermoplastic composites. The principle of welding is that the matrix material locally is heated above the glass transition temperature to bond parts together.

The following welding techniques are currently available:

- ⊙ Resistance welding
- ⊙ Induction welding
- ⊙ Ultrasonic welding

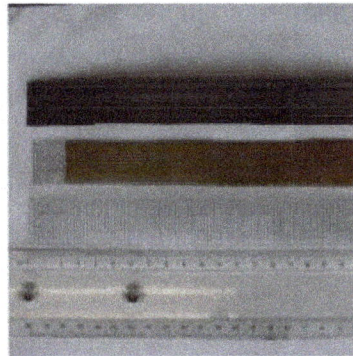

Figure 11.20 Examples of welding in thermoplastics; welding of ribs to the skin of flaps.
(TU Delft, N.D.)

Resistance welding uses the principle of adding electrically resistive material between the surfaces that are welded together. By applying an electrical current through this material, the resistance generated heats up the material at the location of welding. An example application of resistance welding is given in Figure 11.20.

11.6 ADHESIVE BONDING

Adhesive bonding is a permanent joining technique that can only be loaded in shear. Although, adhesive joints exhibit some tensile strength, loading in tension is generally not accepted and avoided. Adhesive bonding is characterised by the high durability. The absence of holes with corresponding stress concentrations makes these joints rather insensitive to fatigue, resulting in fatigue lives of order of magnitude greater than mechanically fastened joints. In addition, because the material is not weakenedduring the bonding process, the adhesive joint is generally stronger than the adherents, i.e. the materials bonded together.

A distinction can be made between hot bonding and cold bonding. The quality of cold bonding is generally lower than hot bonding, because of the adhesives that can be used. As a result, for bonding structural components hot bonding is generally applied. Because of the higher temperature and often higher required pressure autoclaves arenecessary.

Because adhesive bonded joints cannot be inspected for their quality (How do you measure the quality of adhesive after curing?), the process is controlled in detail to guarantee reproducible adhesive joints. Once given process conditions have proven to give bonded joints with sufficient strength, the process can be repeated.

However, the reproducibility of high quality adhesive joints is not yet considered sufficient to assure that such joints will not fail prematurely or at lower stress levels than the failure strength. In addition, the measurement techniques to evaluate the adhesive joint strength non-destructively (without damaging the joint) are non- existent. Therefore, adhesive bonded joints are not allowed in primary structures as single load path solution. Adhesive bonding in these critical structures must always be accompanied with other load transfer paths, such as mechanical fasteners. Bonded stringers contain therefore always rivets at the stringer run-out; the adhesive joint is considered insufficient.

11.6.1 Metallic structures

To create a high quality adhesive bonded joint, the adherents must be pre-treated to provide a good chemical bond. In aluminium structures, the pre-treatment consists of anodising and priming the surfaces before bonding is applied. The anodisation process makes the adhesive joint resistant to corrosion (particularly bond line corrosion), while the primer creates the proper surface to bond the adhesive. These pre-treatment processes require monitoring of the process conditions in order to assure that the pre-treatment layers are applied according to the requirements.

Traditionally, these pre-treatment solutions contain chromates (chromate acid anodising), because of their excellence performance with respect to corrosion resistance. However, because of their detrimental impact on environment, these processes are required to become chromate free. This implies that currently a lot of research is being performed to identify alternative pre-treatment processes that have less impact on the environment, but are as good as the traditional chromate based solutions.

11.6.2 Composite structures

In composite structures the adhesive bonding can be applied in multiple ways. The reason is that the adhesive bonding process requires a curing step. This curing step implied an addition to the metallic production processes, but can be efficiently used in manufacturing of composites, because these materials already require a curing step.

As a consequence, cured composite stringers may be bonded to a composite skin panel that still has to be cured together with the adhesive joint. This combination is often referred to as co-bonding. Alternatively, the skin has been cured in a previous step and the stringers are cured while bonded to the skin at the same time. This combination is commonly referred to as co-curing.

In order to bond two skin panels together, other solutions may be applied that the overlap joint commonly used for metallic panels. The layered structure of the composite skin panel may be utilised to provide a more efficient overlap joint withoutthe geometric step related to the overlap of panels. This solution is often called scarfjoint, as illustrated in Figure 11.21. For example, each layer in the laminate may be terminated at a certain position to provide an overlap with the layer of the other panel. This stepped scarf joint is illustrated in Figure 11.21 (b).

Figure 11.21: Illustration of bonding in composites: scarf joint (a) and stepped scarf joint (b). (Alderlieste 2011)

11.6.3 Adhesive joint strength

The adhesive joint strength relates to the stress distribution throughout the adhesive joint. Although the adhesive joint creates an area of load transfer, rather than locations of load transfer (with rivets in a riveted joint for example), the stresses are not homogeneous.

Consider for example a simple overlap joint as illustrated in Figure 11.22. The primary mode of load transfer is via shear. The adhesive layer will deform by shear as illustrated in the figure. However, because the load is gradually transferred from one sheet to the other, the reduction of stress towards the end of the sheet implies a reduction of strain (or elongation). This is illustrated in the figure by the vertical marks in the sheets. Because the adhesive must deform in a compatible way with the sheets, it implies that peak stress levels occur at the edges of the joint. The shear stress exhibits a so-called bathtub shape.

However, because the two sheets in Figure 11.22 are not aligned, application of a tensile load on the joint will cause the joint to straighten itself. This causes a bending deformation of the sheets with additional tensile stresses in the adhesive, as illustrated in Figure 11.10. These tensile stresses are considered most critical concerning the overall joint strength. Estimating the joint strength therefore implies that not only the shear stresses are to be considered, but also peel stresses as result of these secondary deformation modes.

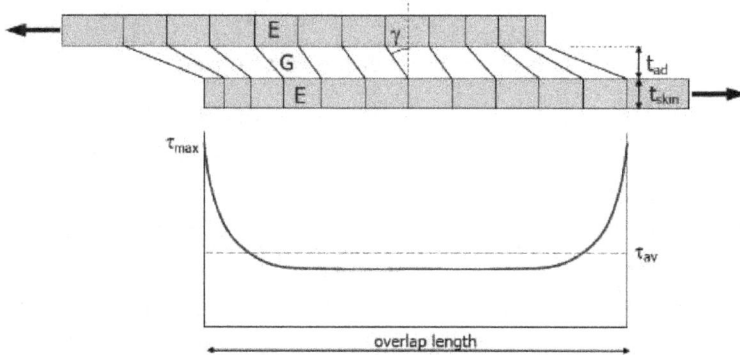

Figure 11.22: Shear stress distribution along the overlap length in a bonded joint. (Alderliesten, 2011)

REFERENCES

CHAPTER 1

Illustrations

Laurensvanlieshout, (2017). Charpy V-notch test.jpg. Retrieved from: https://nl.wikipedia.org/wiki/Bestand: Charpy_V-notch_test.svg, CC-BY-SA 4.0.

NASA, (2003). 86415main_ED03-0180-01_sm.jpg. Retrieved from: https://www.nasa.gov/centers/dryden/news/ResearchUpdate/Helios/Previews/index.html, Public Domain.

NASA, (2003). 86415main_ED03-0180-02_sm.jpg.Retrieved from: https://www.nasa.gov/centers/dryden/news/ResearchUpdate/Helios/Previews/ index. html, Public Domain.

NASA, (2003). 86415main_ED03-0180-03_sm.jpg.Retrieved from: https://www.nasa.gov/centers/dryden/news/ResearchUpdate/Helios/Previews/ index. html, Public Domain.

Otrbski, Andrzej (2013). Proba udarnosci probki 02.jpg. Retrieved from: https://commons.wikimedia.org/wiki/File:Proba_udarnosci_probki_02.jpg. CC-BY-SA 3.0.

CHAPTER 2

Literature

Rice, R.R, Jackson, J.L., Bakuckas, J., and Thompson, S. (2003). Metallic Materials Properties Development and Standardization (Technical report MMPDS-01). U.S. Department of Transportation. Retrieved from: https://ntrl.ntis.gov/NTRL/dashboard/searchResults/titleDetail/PB2003106632.xhtml.

Cohen, A. (2010, May 17). Boeing fixing 787 design flaw – Boeing and Aerospace News [Blog post]. Retrieved from: http://blog.seattlepi.com/aerospace/archives/ 206203.asp.

Illustrations

Anon., (2006). glass-1449943_1920.jpg. Retrieved from https://pixabay.com/en/glass-fire-heat-chemistry-1449943/. CC0.

Asher, Troy – NASA (2014). c-20a_de-iced_iceland_01-31-14.jpg. Retrieved from https://www.nasa.gov/centers/dryden/Features/icelandic_glaciers_study.html.

Public domain.

Britton, Steele C.G. – Senior Airman US Airforce, (2010). 100716-F-7087B-031.jpg. Retrieved from http://www.littlerock.af.mil/News/Article-Display/Article/357068/ fuels-tank-tigers-maintain-the-flow/. Public domain.

Brygg, Simon – Ostersund Photography, (2009). D-AXLG@AJR (3346177101).jpg. Retrieved   from   https://commons.wikimedia.org/wiki/File:D-AXLG_@_AJR_(3346177101).jpg. CC-BY 2.0.

Dahl, Jeff (2007). Jet engine.svg. Retrieved from https://en.wikipedia.org/wiki/File:Jet_engine.svg. CC-BY-SA 4.0.

Deaton, NASA/MSFC, Fred (2013). 0FD8664_400x599.jpg. Retrieved from https://www.nasa.gov/centers/marshall/about/star/star131218.html. Public domain.

Hans, (2012). mountain-bike-23134_1920.jpg. Retrieved from https://pixabay.com/nl/mountainbike-fiets-sneeuwde-23134/. CC0.

Jetstar Airways, (2013). Jetstar's first 787 on the production line (9132370198).jpg. Retrieved   from   https://commons.wikimedia.org/wiki/File:Jetstar%27s_first_787_on_the_production_line_(9132370198).jpg. CC-BY-SA 2.0.

Koul, Anirudh (2008). 2527466246_d8b7c5a75b_z.jpg. Retrieved from https://www.flickr.com/photos/anirudhkoul/2527466246. CC-BY-NC 2.0.

Landis, Tony – NASA (2009). 386210main_ED09-0253-15_full.jpg. Retrieved from https://www.nasa.gov/multimedia/imagegallery/image_feature_1469.html. Public Domain.

NASA, (2006). 146600main_sts1anniv-AC76-1713.jpg. Retrieved from https://www.nasa.gov/centers/ames/images/content/146600main_sts1anniv-

AC76-1713.jpg. Public Domain.

NASA-Goddard Space Flight Center, (2011). cnofs-orig.jpg. Retrieved from https://www.nasa.gov/mission_pages/sunearth/news/lightning-waves.html. Public Domain.

NTSB, (2010). Center Wing Fuel Tank.png. Retrieved from https://commons. wikimedia.org/wiki/File:Center_Wing_Fuel_Tank.png.PublicDomain.

OU Department of Materials Engineering, (2004). ccf4.jpg. Retrieved from http:// pardo.net/bike/pic/fail-009/000.html. Copyright OU.

SAC Stier, James (2009). 800px-Merlin_Helicopter_in_Californian_Desert_During_ Ex_Merlin_Vortex_MOD_45150792.j pg. Retrieved from https://commons. wikimedia.org/wiki/ File:Merlin_Helicopter_in_Californian_Desert_During_ Ex_Merlin_Vortex_MOD_451507 92.jpg. Open Government Licence.

Schida, Tom – NASA, (2005). 274858main_EC05-0166-21_full_full.jpg. Retrieved from   https://www.nasa.gov/centers/dryden/multimedia/imagegallery/ Shuttle/ EC05-0166-21.html. Public Domain.

Skeeze, (2008). helocasting-704447_1920.jpg. Retrieved from https://pixabay.com/nl/ helocasting-helikopter-water-704447/, CC0.

U.S. GPO, (1943). TankerSchenectady.jpg. Retrieved from https://nl.wikipedia.org/ wiki/Bestand:TankerSchenectady.jpg. Public Domain.

Wollman, Gina K. – Mass Communication Specialist 2nd Class U.S. Navy, (2010). US_Navy_100612-N-3595W-004_Sailors_assigned_to_the_Patriots_of_ Electronic_Att ack_Squadron_(VAQ)_140,_clean_an_EA-6B_Prowler_on_the_ flight_deck_of_the_aircr aft_carrier_USS_Dwight_D._Eisenhower_(CVN_69). jpg. Retrieved fromhttps://commons.wikimedia.org/wiki/

File:US_Navy_100612-N-3595W-004_Sailors_assigned_to_the_Patriots_of_ Electronic_ Attack_Squadron_(VAQ)_140,_clean_an_EA-6B_Prowler_on_the_ flight_deck_of_the_air craft_carrier_USS_Dwight_D._Eisenhower_(CVN_69). jpg. Public Domain.

CHAPTER 3

Literature

Vlot, A., & Gunnink, J. W. (2001). Fibre Metal Laminates: An Introduction. Dordrecht: Springer The Netherlands.

Illustrations

Atkeison, Charles, (2003). shuttlenExplorer starboard nose side 2003.jpg. Retrieved from   https://commons.m.wikimedia.org/wiki/   File:Shuttle_Explorer_ starboard_nose_side_2003.jpg#mw-jump-to-license. CC-BY- SA2.0.

Anon.,(2017).wc_bathroom_toilet_ceramics_hygiene_discharge_vater_ process-1059556.jpg!d.jpg.   Retrieved   from   https://pxhere.com/en/ photo/1059556. CC0.

Bender, Ben, (2014). Kop van Zuid, Rotterdam, Netherlands – panoramio (12).jpg. Retrieved from https://commons.wikimedia.org/wiki/ File:Kop_van_Zuid,_ Rotterdam,_Netherlands_-_panoramio_(12).jpg?uselang=nl. CC- BY-SA 3.0.

Boffoli, CJ, (2018). 787fuselage.jpg. Retrieved from https://commons.wikimedia. org/wiki/File:787fuselage.jpg. Public domain.

Cjp24, (2009). Glass reinforcements.jpg.

Retrieved from https://commons.wikimedia.org/wiki/File:Glass_reinforcements. jpg#/media/ File:Glass_reinforcements.jpg. CC-BY-SA 3.0.

Gnokii, (2011). Tennis-rack.svg. Retrieved from https://commons.wikimedia.org/ wiki/File:Tennis-rack.svg. CC0.

GuentherDilingen,   (2012). air-sports-2199683_960_720.jpg.   Retrieved fromhttps://pixabay.com/en/air-sports-motorsegler-pilot-2199683/. CC0.

Hans, (2013). dustbin-95181_1920.jpg. Retrieved from https://pixabay.com/ get/   e030b00729e9002ad25a5840981318c3fe76e7d71bb9154094f2c4/ dustbin-95181_1920.jpg. CC0.

KarinKarin, (2015). ice-684741_960_720.jpg. Retrieved from https://pixabay.com/ en/ice-train-railway-express-train-684741/. Public Domain.

Medienluemmel, (2016).   performance-3120640_960_720.jpg. Retrieved from https://pixabay.com/en/performance-turbine-wind-3120640/. CC0.

NASA, (2003a). CAIB_highres_full.pdf:figure3-1-1.   Retrieved from https://www. nasa.gov/columbia/home/CAIB_Vol1.html. Public Domain.

NASA, (2003b). CAIB_highres_full.pdfFigure2-3-2.jpg. Retrieved from https:// www.nasa.gov/columbia/home/CAIB_Vol1.html. Public Domain.

NASA, (2003c). CAIB_highres_full.pdfFigure2-6-1.   Retrieved from https://www. nasa.gov/columbia/home/CAIB_Vol1.html. Public Domain.

NASA, (2003d). CAIB_highres_full.pdfFigure 3-8-9. Retrieved from https://www. nasa.gov/columbia/home/CAIB_Vol1.html. Public Domain.

NASA, (2003e). CAIB_highres_full.pdfFigure3-8-5. Retrieved from https://www. nasa.gov/columbia/home/CAIB_Vol1.html. Public Domain.

NASA-STS-118 crew,   (2007).

Damaged_TPS_Tiles_of_Endeavour_(NASA_S118-E-06229).jpg.   Retrieved   from https://commons.wikimedia.org/wiki/   File:Damaged_TPS_Tiles_of_ Endeavour_(NASA_S118-E-06229).jpg. Public Domain.

NI-CO-LE,(2017).plant-production-2335621_1920pg.Retrievedfromhttps://pixabay. com/get/ eb36b20a2ef6003ed1534705fb0938c9bd22ffd41cb3134996f9c47aa2/ plant- production-2335621_1920.jpg?attachment. CC0.

Pexels, (2016). pile-1868894_960_720.jpg. Retrieved from https://pixabay.com/en/pile-rubber-stacked-tires-1868894/. CC0.

Pingstone, Adrian, (2004). American.airlines.b777.arp.jpg. Retrieved from https://commons.wikimedia.org/wiki/File:American.airlines.b777.arp.jpg. Public Domain.

PMullahaa, (2015). NIJ_LVLIIIA_Kogelvrij_vest,_BA8001.jpg. Retrieved from https://commons.wikimedia.org/w/index.php?curid=37956474. CC-BY-SA 3.0.

Torr68, (2005). 1280px-GF-04_b.jpg. Retrieved from https://commons.wikimedia.org/wiki/File:GF-04_b.JPG#/media/File:GF-04_b.JPG. CC-BY-SA 3.0.

Vinayr16, (2014). tire-2645692_960_720.jpg. Retrieved from https://pixabay.com/en/tire-tread-new-wheel-rubber-car-2645692/. CC0.

Volk, Willy, (2008). 2536399954_61b8770f95_z.jpg. Retrieved from https://www.flickr.com/photos/volk/2536399954. CC-BY-NC-SA 2.0.

Yogipurnama, (2017). lego-2200009_960_720.jpg. Retrieved from https://pixabay.com/en/lego-toys-kids-aircraft-helicopter-2200009/, CC0.

CHAPTER 4

Illustrations

Arnd, Florian, (2005). Strangziehverfahren.png. Retrieved from https://commons.wikimedia.org/wiki/File:Strangziehverfahren.png. CC-BY-SA 3.0.

CM_Foto, (2016). cylinder-1369348_960_720.jpg. Retrieved from https://pixabay.com/en/cylinder-engine-block-motor-1369348/. CC0.

Contest Group, (2016). 4-10-a.jpg. Retrieved from www.contestyachts.com. Copyright by Contest Group. Used with permission.

Esi.us1, (2010). Enrollado-de-filamentos_1.jpg. Retrieved from https://commons.wikimedia.org/wiki/File:Enrollado-de-filamentos_1.jpg. PublicDomain.

Gdipasquale1, (2018). Filam.jpg. Retrieved from https://commons.wikimedia.org/wiki/File:Filam.jpg. CC-BY-SA 4.0.

Knechtel – Donald, C. – Airman 1st Class U.S. Airforce, (2017). 170222-F-SE307-086.JPG. Retrieved from https://www.afgsc.af.mil/News/Photos/igphoto/2001706654/. Public Domain.

Lightweight Structures BV, (2006). 4-10-b.jpg. Retrieved from http://www.lightweight-structures.com/. Copyright by Lightweight Structures BV. Used with permission.

Lightweight Structures BV, (2006). 4-10-c.jpg. Retrieved from http://www.lightweight-structures.com/. Copyright by Lightweight Structures BV. Used with permission.

MarPockStudios, (2017). crucible-2109202_960_720.jpg. Retrieved from https://pixabay.com/en/crucible-foundry-molten-bronze-2109202/. CC0.

CHAPTER 5

Literature

Rawal, S. (January 01, 2001). Metal-Matrix Composites for Space Applications. JOM Warrendale-, 53, 14-17.

Wertz, J. R., & Larson, W. J. (1999). Space mission analysis and design. Torrance, Calif: Microcosm Press/Dordrecht: Kluwer Academic Publishers, The Netherlands.

Illustrations

Altoing, (2011), Su25-kompo-vers2.svg. Retrieved fromhttps://commons.wikimedia.org/wiki/File:Su25-kompo-vers2.svg. CC-BY-SA 3.0.

Anon., (n.d.), 13_53_1.jpg.Retrieved from http://www.tpub.com/air/13-42.htm via AVIATION STRUCTURAL MECHANIC E1 & C. Public Domain.

Anon., (n.d.), 5-1.jpg. Retrieved from Lecture Notes TU Delft, AE1102 Introduction into Aerospace Structures and Materials, Alderliesten, 2011 (hardcopy). Public Domain.

Anon., (n.d.), 5-27-b.jpg. Retrieved from Lecture Notes TU Delft, AE1102 Introduction into Aerospace Structures and Materials, Alderliesten, 2011 (hardcopy). Public domain.

Anon., (n.d.), 5-36-c.jpg.Retrieved from Lecture Notes TU Delft, AE1102 Introduction into Aerospace Structures and Materials, Alderliesten, 2011 (hardcopy). Public Domain.

Bidini, Aldo, (2012), Dassault-Breguet_Atlantique_2,_France_-_Navy_JP7415097.jpg. Retrieved from https://commons.wikimedia.org/wiki/File:Dassault-Breguet_Atlantique_2,_France_-_Navy_JP7415097.jpg. Public Domain.

Dr Brains, (2013), Airbus Lagardère P45 -Wing equipping (MSN120) (1). JPG.Retrievedfrom https://commons.wikimedia.org/wiki/File:Airbus_Lagard%C3%A8re_P45_-_Wing_equipping_(MSN120)_(1).JPG. CC0.

ESA, (2012), MSG-3_In_the_cleanroom_node_full_image_2. Retrieved from https://www.esa.int/spaceinimages/Images/2012/03/MSG-3_In_the_cleanroom.

Copyright ESA.

ESA, (2002), 5-38-a.jpg.Retrieved from https://esa.int. Copyright ESA.

ESA, (n.d.), 5-38-b.jpg.Retrieved from https://esa.int. Copyright ESA.

ESA, (2002),

De_verschillende_instrumenten_aan_boord_van_Envisat_node_full_image_2. Retrieved from http://www.esa.int/spaceinimages/Images/2002/07/ De_verschillende_instrumenten_aan_boord_van_Envisat. Copyright ESA.

Kolossos, (2006), Fuselage-747.jpg.Retrieved from https://commons.wikimedia. org/wiki/File:Fuselage-747.jpg?uselang=en. CC-BY-SA 3.0.

Le Floc'h – ESA, Anneke, (2002),Envisat_Structural_Model_on_the_HYDRA_Shaker_ in_the_ESTEC_Tests_Centre_Fl001

_node_full_image_2.jpg. Retrieved from https://www.esa.int/spaceinimages/ Images/ 2001/11/

Envisat_Structural_Model_on_the_HYDRA_Shaker_in_the_ESTEC_Tests_Centre_ Fl001 . Copyright ESA.

NASA, (2000), boeing16.jpg. Retrieved from https://stardust.jpl.nasa.gov/mission/ rockettour.html. Public Domain.

NASA, (2001), fig1-sm.jpg. Retrieved from https://www.tms.org/pubs/journals/ JOM/0104/Rawal-0104.html. Public Domain.

NASA, (2007), 221640main_EDC_Spacecraft_Structures.pdf:Figure 6.3. Retrieved from https://www.nasa.gov/pdf/221640main_EDC_Spacecraft_Structures. pdf. PublicDomain.

NASA, (2007), 221640main_EDC_Spacecraft_Structures.pdf:Figure 6.4. Retrieved from https://www.nasa.gov/pdf/221640main_EDC_Spacecraft_Structures. pdf. PublicDomain.

NASA, (2010), 20100005273.pdf. Retrieved from https://ntrs.nasa.gov/archive/ nasa/casi.ntrs.nasa.gov/20100005273.pdf. Public Domain.

NASA, (n.d.), Second_Stage.pdf. Retrieved from https://history.msfc.nasa.gov/ saturn_apollo/documents/Second_Stage.pdf. Public Domain.

NASA, (n.d.), p197.jpg. Retrieved from https://history.nasa.gov/SP-4206/p197. htm. Public Domain.

NASA, (n.d.), 5-37-a.jpg. Retrieved from https://nasa.gov. Public Domain. NASA, (n.d.), 5-37-b.jpg. Retrieved from https://nasa.gov. Public Domain.

Transportation Safety Board of Canada, (2007), a05f0047.pdf:Photo 3. Retrieved from http://www.tsb.gc.ca/eng/rapports-reports/aviation/2005/a05f0047/ a05f0047.pdf. Copyright TSB.

Transportation Safety Board of Canada, (2007), a05f0047.pdf: Figure 4. Retrieved from http://www.tsb.gc.ca/eng/rapports-reports/aviation/2005/a05f0047/ a05f0047.pdf. Copyright TSB.

Transportation Safety Board of Canada, (2007), a05f0047.pdf:Photo 3. Retrieved from http://www.tsb.gc.ca/eng/rapports-reports/aviation/2005/a05f0047/a05f0047.pdf. Copyright TSB.

Transportation Safety Board of Canada, (2007), 5-14-f.jpg. Retrieved from http://www.tsb.gc.ca/eng/rapports-reports/aviation/2005/a05f0047/a05f0047.pdf. Copyright TSB.

Sapp – U.S. Air Force, Sue, (2007), 071130-F-5350S-205.JPG .Retrieved from http://www.af.mil/News/Photos/igphoto/2000419184/. Public Domain.

CHAPTER 6

Illustrations

Ienac, (2006), A_380_meeting.jpg. Retrieved from https://commons.wikimedia.org/ wiki/File:A_380_meeting.jpg. Public Domain.

Lämpel, (2013), Airbus_A380_on_MAKS_2011_crop.jpg. Retrieved from https://commons.wikimedia.org/wiki/File:Airbus_A380_on_MAKS_2011_crop.jpg.

CC-BY-SA 3.0.

Mr_worker, (2014), airbus-2466266_960_720.jpg. Retrieved from https://pixabay.com/en/airbus-a380-frankfurt-start-flight-2466266/. Public Domain.

NASA, (2017), 6-17.jpg. Retrieved from https://NASA.gov. Public Domain.

NASA, (2000), KSC-00PD-1263.jpg.jpg. Retrieved from https://science.ksc.nasa.gov/shuttle/missions/sts-106/images/high/KSC-00PD-1263.jpg. Public Domain.

NASA, (n.d.), 6-18-b.jpg. Retrieved from https://nasa.gov. Public Domain.

Noret, Phillipe – AirTeamimages, (2012), FGSPN.jpg. Retrieved from https://commons.wikimedia.org/wiki/File:FGSPN.jpg. CC-BY-SA 2.5.

Roddy, K.–UC Davis, (2007), buttress.jpeg. Retrieved fromhttp://medieval.ucdavis.edu/20B/buttress.jpeg.

Roddy, K. – UC Davis, (2007), choir.jpeg. Retrieved from http://medieval.ucdavis.edu/20B/choir.jpeg.

Sapp, Sue – U.S. Air Force, (2011), 110427-F-EL616-189.JPG. Retrieved from http://www.robins.af.mil/News/Photos/igphoto/2000252075/. Public Domain.

Sapp, Sue – U.S. Air Force, (2006), 061115-F-2383G-109.jpeg. Retrieved from http://www.afmc.af.mil/News/Photos/igphoto/2000534465/. Public Domain.

CHAPTER 8

Literature

Rice, R.R, Jackson, J.L., Bakuckas, J., and Thompson, S. (2003) Metallic Materials Properties Development and Standardization (Technical report MMPDS-01). U.S. Department of Transportation. Retrieved from: https://ntrl.ntis.gov/NTRL/dashboard/searchResults/titleDetail/PB2003106632.xhtml

Wertz, J. R., & Larson, W. J. (1999). Space mission analysis and design. Torrance, Calif: Microcosm Press/Dordrecht: Kluwer Academic Publishers, The Netherlands.

Illustrations

Cliff, (2008), 3348955799_907d31c689_o.jpg. Retrieved from https://www.flickr.com/photos/nostri-imago/3348955799. CC-BY 2.0.

Fietstijden.nl, (2018), frame.jpg. Retrieved from https://www.fietstijden.nl/wielrennen/onderdelen/frame. Public Domain.

Glorycycles, (2018), 24730908477_1804e4c2ec_o.jpg. Retrieved from https://www.flickr.com/photos/glorycycles/24730908477/in/photolist-

DFontK-22fABzY-DFompk-22fAAuS-23kbcuk-23kbc5n-DjqdeK-Y7fn1i-YjFcXz-XgmMm5-YjFbQz-26ufLm6-26phCf3-26i55x4-Ja92Lh-25WFPXN-23PK4v4-25aiNgp- 25adQrr-25addXF-259AhmM-255vXcY-Ft8inz-

GZqf3s-254ndH1-GZqeaW-23L154T-22n6zCo-23L14jM-22n6yVG-23L13oZ-23L12ue-

22n6xNw-2432kGC-22bJhaj-23zzzpX-GNWJvW-

FhCatM-24WF5r6-FhC9H8-23RYEkq-23RYDVs-23RYDB1-23RYDmS-23RYD1w-23RY CEw-23RYCmL-23RYBUy-23RWPb5-EDb9cn. CC-BY 2.0.

Kolossos, (2006), Fuselage-747.jpg, Retrieved from https://commons.wikimedia.org/wiki/File:Fuselage-747.jpg?uselang=en. CC-BY-SA 3.0.

Saunders, Kevin G., (2008), 4635297991_1f275c6865_o.jpg. Retrieved from https://www.flickr.com/photos/kgsbikes/4635297991. CC-BY-NC-ND2.0.

steel-vintage.com, (2018), bianchi-specialissima-road-bicycle-frame-1.jpg. Retrieved from https://www.steel-vintage.com/sold-bikes-and-parts/57/?limit=60, CC-BY.

Turner, Roland, (2012), Ariane_5,_Musee_de_l%27Air_et_de_l%27Espace,_Le_Bourget,_Paris._(82089 75545).jpg. Retrieved from https://commons.wikimedia.org/wiki/ File:Ariane_5,_Musee_de_l%27Air_et_de_l%27Espace,_Le_Bourget,_Paris._(82089755 45).jpg. CC-BY-SA2.0.

CHAPTER 9

Literature

Beck, Laurie F, Dellinger, Ann M., O'Neil, Mary E. (2007, 15 July). Motor Vehicle Crash Injury Rates by Mode of Travel, United States: Using Exposure-Based Methods to Quantify Differences, American Journal of Epidemiology, Volume 166, Issue 2, Pages212–218, https://doi.org/10.1093/aje/kwm064.

Ford, R. (2000, October) The Risks of Travel, Modern Railways, no. 10.

International Civil Aviation Organization. (2008). Annual reports of the Council. Montreal: ICAO. Retrieved from: https://www.icao.int/publications/ Documents/ 9916_en.pdf#search=doc%209916.

Schijve, J. (2010). Fatigue of structures and materials. Dordrecht: Springer, The Netherlands.

Illustrations

British Airways , (1952). CM1mk1galyp.jpg. Retrieved f r o m https://www.fose1.plymouth.ac.uk/fatiguefracture/tutorials/FailureCases/ images/ CM1mk1galyp.jpg. Public Domain.

Dtom, (2007). Wing_structure1.JPG. Retrieved from https://commons.wikimedia. org/ wiki/File:Wing_structure1.JPG#/media/File:Wing_structure1.JPG. Public Domain.

Fokker, (n.d.). 9-3.jpg. Retrieved from Lecture Notes TU Delft, AE1102 Introduction into Aerospace Structures and Materials, Alderliesten, 2011 (hardcopy). Public Domain

Krelnik, (2009). Fuselage_of_de_Havilland_Comet_Airliner_G-ALYP.JPG. Retrieved from https://en.wikipedia.org/wiki/File:Fuselage_of_de_Havilland_Comet_ Airliner_G- ALYP.JPG. CC-BY-SA 3.0.

Ministry of transport and civil aviation, (1954). G-ALYP.pdf:Figure 2. Retrieved from https://www.baaa-acro.com/sites/default/files/import/uploads/2017/04/ G-ALYP.pdf. Public Domain.

National Transportation Safety Board, (1988), 9-7-a.jpg. Retrieved from https:// www.ntsb.gov/Pages/default.aspx. Public Domain.

National Transportation Safety Board, (1988). AAR89-03.pdf:Figure2b. Retrieved from https://www.ntsb.gov/investigations/AccidentReports/Reports/ AAR8903.pdf. Public Domain.

National Transportation Safety Board, (1988). 9-7-c.jpg. Retrieved from https:// www.ntsb.gov/Pages/default.aspx. Public Domain

Ulbrich,   Ken   –   NASA, (2014). :NASA's_G-
III_flying_test_bed_aircraft_rests_on_three_pneumatic_lifting_devices_
  or_"airbags"_in_ preparation_for_loads_testing.jpg. Retrieved from https://
  commons.wikimedia.org/ wiki/File:NASA's_G- III_flying_test_bed_aircraft_
  rests_on_three_pneumatic_lifting_devices_or_%22airbags

%22_in_preparation_for_loads_testing.jpg#/media/File:NASA's_G- III_flying_test_
  bed_aircraft_rests_on_three_pneumatic_lifting_devices_or_"airbags"_in_
  preparation_for_loads_testing.jpg. Public Domain.

Wikipedia , (2004). 9-1.jpg. Retrieved from   https://en.wikipedia.org/wiki/
  Aviation_safety. CC-BY-SA3.0.

Wizard191, (2010). ensile_specimen-round_and_flat.jpg. Retrieved from https://
  commons.wikimedia.org/wiki/File:Tensile_specimen-round_and_flat.jpg. CC-
  BY-SA3.0.

Yapparina, (2014).   Crack_in_a_finite_plate_under_uniform_uniaxial_stress.
  png. Retrieved from https://commons.wikimedia.org/wiki/File:Crack_in_a_
  finite_plate_under_uniform_uniaxial_stress.png. CC-BY-SA3.0.

CHAPTER 10

Literature

Pilkey, W. (1997). Peterson`s stress concentration factors. Second edition. John
  Wiley & Sons, Canada.

Schijve, J. (2010). Fatigue of structures and materials. Dordrecht: Springer, The
  Netherlands.

Illustrations

Anon., (n.d.). rankine.jpg. Retrieved from: http://www.daviddarling.info/
  encyclopedia/R/Rankine.html. Public Domain.

Antiproton, (2017). grains-2163690_1920.jpg. Retrieved from https://pixabay.com/
  en/grains-metal-alloy-microscope-2163690/. CC0.

Carnevale,   Paola, (2014). PaolaCarnevale.thesis.pdf.   Retrieved   f r o m
  https://repository.tudelft.nl/islandora/object/uuid%3A880605ea-fb4c-
  4b9b-a25e- ad610d94a5cd. Copyright Carnevale. Used with Permission.

Glynn, Joseph,(1844). Tender_fatigued_axle.JPG.   Retrieved   from https://
  en.wikipedia.org/wiki/File:Tender_fatigued_axle.JPG . Public Domain.

MT Aerospace   AG,   (2006). CVIpullout.jpg.   Retrieved   f r o m
  https://commons.wikimedia.org/wiki/File:CVIpullout.jpg. CC-BY-SA3.0.

Plymouth Electron Microscopy Centre (PEMC), (n.d.). 10-16-c.jpg. Retrieved from https://www.fose1.plymouth.ac.uk/sme/MATS347/MATS347A4%20 fracture.htm . Copyright PEMC. Used with Permission.

Plymouth Electron Microscopy Centre (PEMC), (n.d.). 10-16-d.jpg. Retrieved from https://www.fose1.plymouth.ac.uk/sme/MATS347/MATS347A4%20 fracture.htm . Copyright PEMC. Used with Permission.

Unknown Artist, (1842). Meudon_1842.jpg. Retrieved from https:// commons.wikimedia.org/wiki/File:Meudon_1842.jpg. Public Domain.

Unknown Artist, (1842). 10-9-a.jpg. Retrieved from Lecture Notes TU Delft, AE1102 Introduction into Aerospace Structures and Materials, Alderliesten, 2011 (hardcopy). Public Domain.

CHAPTER 11

Illustrations

Alexas_Fotos, (2016). screw-1711469_1920.jpg. Retrieved from https://pixabay. com/en/screw-nuts-hex-bolt-1711469/. Public Domain.

Cdang, (2010). Percages_et_vis.JPG. Retrieved from https://commons. wikimedia.org/wiki/File:Percages_et_vis.JPG. Public Domain.

InspiredImages, (2015). metal-950169_1920.jpg. Retrieved from https://pixabay. com/nl/metaal-nagels-staal-metalen-950169/. CC0

Schijve, J. (2010). 11-12.jpg. Retrieved from Schijve, J. (2010). Fatigue of structures and materials. Dordrecht: Springer, The Netherlands (hardcopy). Copyright Schijve. Used with permission.

DEFINITIONS

Airframe	Structure that takes up all forces during operation of the vehicle.
Anisotropic material	Material with mechanical properties depending on the direction
Brittileness	Property of material that allows little bending or deformation without shattering
Composites	Engineering materials containing two or more distinct and structurally complementary substances with different physical or chemical properties, having structural or functional properties not present in the individual substances
Contraction/ Expansion	Reaction produced in material as the result of heating or In general; degradation of engineering materials due to
Corrosion:	chemical reaction with its environment. Metal specific: electrochemical oxidation in reaction with an oxidant such as oxygen.
Density	Weight of a unit volume of material
Ductility	Property of metal that allows it to be permanently drawn, bent, or twisted into various shapes without breaking
Elasticity	Property enables material to return to its original shape when the force which causes the change of shape is removed
Engineering stress	Load divided by the original cross-section of the material/ component.
Fatigue	Damage phenomenon induced by multiple load cycles below ultimate strength of material or structure causing permanent deterioration of material or structure resulting in a reduction in load bearing capability
Fracture toughness	Parameter describing the resistance of a material against fracture in presence of cracks in tensile mode
Galvanic corrosion	Electrochemical process in which one metal corrodes when in electrical contact with a different type of metal or material and both metals are immersed in a substance containing electrolyte.

Geometrical tolerances	Maximum variation allowed in form or positioning
Glass transition temperature	The temperature at which polymers exhibit a transition from a more glassy (hard or brittle) state to a rubbery (elastic or flexible) state Ability to resist abrasion, penetration, cutting or permanent distortion
Isotropic material	Material having identical mechanical properties in all directions
Primary structural element	Critical load bearing structure of an aircraft/spacecraft that in case of severe damage will fail the entire aircraft/spacecraft
Quasi-isotropic material:	Approximation of isotropic material by placing multiple anisotropic layers in different directions
RTM	Resin Transfer Moulding
Saint Vanant's Principle	Disturbances in the stress field remain limited to the direct neighbourhood of the location of disturbance
Secondary structural element	Structural elements of an aircraft/spacecraft that carry only air and inertial loads generated on or in the secondary structure
Specific property	Material property divided by its density
Stress concentration factor	Parameter describing the relation between the peak stress at the (blunt) notch root and the nominal (nett) stress in the cross-section
Stress intensity factor	Parameter describing the severity (intensity) of stresses near the crack tip
Toughness	Property of a material that withstands tearing or shearing and may be stretched without being deformed or breaking
True stress	Load divided by the actual cross-section of the material/
VARTM	Vacuum Assisted Resin Transfer Moulding
VI	Vacuum Infusion

INDEX

www.ingramcontent.com/pod-product-compliance
Lightning Source LLC
Chambersburg PA
CBHW062001190326
41458CB00009B/2935